THE ORIGINS OF OPEN-FIELD AGRICULTURE

The Origins of
Open-Field Agriculture

Edited by Trevor Rowley

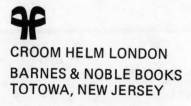

CROOM HELM LONDON
BARNES & NOBLE BOOKS
TOTOWA, NEW JERSEY

© 1981 T. Rowley
Croom Helm Ltd, 2-10 St John's Road, London SW11

British Library Cataloguing in Publication Data

The origins of open-field agriculture. — (Croom Helm historical geography series)
    1. Land tenure — Great Britain — History
    2. Agricultural systems — Great Britain — History
    I. Title   II. Rowley, Trevor
    333.2   HD594   80-41574

    ISBN 0-7099-0170-4

First published in the USA 1981 by
BARNES & NOBLE BOOKS
81 ADAMS DRIVE
TOTOWA, New Jersey, 07512
ISBN: 0-389-20102-2

Typesetting by Elephant Productions, London
Printed and bound in Great Britain by
Biddles Ltd, Guildford and King's Lynn

CONTENTS

FIGURES AND PLATES

Figures

Figures

Plates

PREFACE

The papers in this book were presented to a seminar on the Origins of Open-field Agriculture held at the Oxford University Department for External Studies in November 1978. They formed part of the programme to mark 100 years of adult education in Oxford University.

The aim of holding the seminar, at which the majority of these papers were presented, was to bring together scholars of different disciplines to explore their current attitudes to the origins and early development of open-field agriculture in England and Wales. It was decided from the outset to use the term 'open fields' rather than medieval, strip or common fields as it seemed to be the most objective terminology. The others all carried with them certain implications about chronology, tenure or form. Despite their very different backgrounds, one thing all the contributors agreed about was that open fields varied enormously both in time and in space and that there were several regional forms of open fields.

Open-field agriculture has provided a fertile area for debate over the past century. In recent years questions about the way in which the open fields originated have become increasingly pertinent, as traditional ideas about their pagan-Saxon beginnings have been discarded. Scholars no longer accept that the change from prehistoric Celtic fields to strip cultivation took place in the post-Roman migration period. Moreover, it is now increasingly believed that this fundamental change in agricultural organisation may have been associated with the process of settlement nucleation in the later Anglo-Saxon period.

Taylor, who opened the proceedings, was pessimistic about the contribution archaeology could make to understanding the origins of the system. He pointed out that archaeological techniques would only be able to identify incidents of open-field farming when it was already well established and that much of our available field evidence in the form of earthworks was late or even post-medieval in date. However, Hall and Barker and Higham suggested ingenious ways in which a combination of judicious fieldwork and selective excavation may well produce evidence, if not of the origins of open-field systems then of open fields operating at an early date. Barker and Higham, whose paper is not included in this collection, discussed the excavated pre-Norman ridge and furrow found underneath the motte and bailey castle at Hen

Domen, and the earthworks of apparently similar field systems found in areas adjacent to the castle.

The place-name historians, represented in this volume by Hooke, demonstrate ways in which careful analysis of place-name evidence can be used to help understand aspects of early field systems. Fox provides a detailed account of the Midland system, its variations and examples of field rearrangement.

A more recent conventional wisdom is questioned by Campbell who argues that far from reflecting an increasing population in the twelfth and thirteenth centuries the development of open-field farming was more likely to be the result of a declining population. He also suggests a more analytical approach to the problem.

The remainder of the papers in the volume incorporate a similar variety of views reflecting either regional forms of open-field system or conflicting suggestions about the way, the date or method in which they evolved. It is impossible to come to any conclusions which would be either chronologically or geographically valid. There was no consensus at the seminar. The nature of these papers does however indicate that the subject is a live one and one into which an enormous amount of fruitful research is currently taking place. It is hoped that the papers in this volume will provide a stimulus to scholars from all the disciplines that are involved in the exploration of the origins of open-field agriculture.

Thanks are due to Shirley Hermon, Lucienne Walker and Linda Rowley for helping with the production of the typescript and BKS Surveys for permission to produce Plate 2. I would also like to thank contributors for their work in producing the papers.

Trevor Rowley

1 ARCHAEOLOGY AND THE ORIGINS OF OPEN-FIELD AGRICULTURE

C.C. Taylor

Numerous scholars have attempted to describe, define and explain the general system of agriculture practised over much of Britain, and indeed on the continent, throughout most of the medieval period and later. However, in spite of all the effort put into the study there still seems little agreement on what constituted open-field agriculture and how and when it developed. At its simplest, open-field agriculture was the means by which land was cultivated by the inhabitants of a township who worked their holdings in unenclosed parcels.

This definition hardly explains the true complexity or variety of the system as it is revealed through medieval and post-medieval documentation. For the definition not only embraces the highly developed Midland, or regular, common field system as first defined by Gray as well as the numerous other variants described in detail by Dr Campbell (see below pp. 112-29), but also covers other related systems such as runrig and infield and outfield, which are well documented as having existed in many places in this country. All these variants can be, and indeed have been, defined to a greater or lesser extent, but from the archaeological point of view it is important to note the main characteristics which separate one type from another. Dr Thirsk defined the classic Midland common field system as being made up of four essential elements (Thirsk, 1966). First, arable and meadow were divided into strips among the cultivators. Secondly, both arable and meadow were thrown open for common pasturing at certain times. Thirdly, there were common rights over waste and lastly the ordering of these activities was regulated by some form of assembly of the people involved. On the other hand the other more irregular common field systems are differentiated by such features as whether cropping or pasturing is undertaken in common or whether regulation is by individuals or groups. All these and many other factors such as crop rotations, tenurial arrangements and inheritance laws have been used to explain the many different forms that open-field agriculture is known to have taken. In the final analysis, an assessment of the relative importance of the different factors which go to make up the various types of open-field systems might go some way towards explaining the origins of such systems.

All these definitions and factors, however, are a long way from the archaeologists' data base for open-field agriculture. Archaeologists, by the very nature of their discipline, are concerned with the material remains of the past, and thus they are mainly involved with the physical manifestations of agricultural systems, for example ridge-and-furrow, embanked strips and strip lynchets.

At once we are faced with an almost insuperable problem. Most of the definitions of the various types of open fields, admirable though they are, usually say nothing about the only aspect of open-field agriculture that the archaeologist can grasp, the physical remains of cultivation. The closest that archaeologists can approach to most of the definitions is through the use of the word strips. The strips, as defined by the historians or geographers, are the basic units of tenure and cultivation; the archaeologist's ridge-and-furrow or strip lynchets might be considered to be related. They clearly are related by the end of the medieval period, at least to some extent, but this is not to say that they were always so linked. Indeed if it is possible to criticise most of the definitions of open-field agriculture it is in their inclusion of the word strip. It is at least theoretically possible that open-field arable could be divided among the cultivators in blocks of land of very different shapes. All that would be necessary is for the cultivators to have a different type of plough and/or to use a different form of ploughing technique from those that produce ridge-and-furrow or long strips.

Here lies the basic difference between the historical approach and that of the archaeologist. The historians' and the historical geographers' definitions of open fields, brought out in the other papers of this volume, are definitions not merely of a method of cultivation but of systems of cropping, tenure and social organisation. This is far removed from the relatively simple remains of plough ridges, pottery and buried soils which are the concern of the archaeologists. This is not to say that the archaeologists have no part to play in the understanding of the origins of open fields. They certainly have, but it must be recognised that they are working from a standpoint very different from that of the historians and much of what is recoverable from documentary sources is often unrelated to the evidence that archaeologists can provide.

The difference between archaeologists and historians studying open fields is rather like two parallel railway tracks. Most of the time the tracks are together but in a slightly different place. Occasionally there are points where the tracks join and there are numerous branch lines leading off in various directions. Both tracks are heading for the same terminus, but one suspects that when they reach it they will end on adjacent platforms. Both archaeologists and historians must be very conscious of the different forms of evidence they are dealing with in their common study of open fields. Most of the evidence is not directly related and some of it may even conflict. It is not that one form of evidence is better or worse than the other, it is that they are different and thus tell us about different aspects of the same phenomena.

With this basic thought in our minds, let us look at the archaeological problems in discovering the origins of open fields. First, a general point: it is extremely difficult to discover the origins of almost any aspect of human behaviour, for until a feature, technique or organisation is relatively common-place and well developed, it is usually very difficult to detect archaeologically. Thus on general grounds it is not easy for archaeologists to find the origin of a system of cultivation, even in simple terms, regardless of the other and more important aspects of the definitions of open fields.

It is also necessary to consider what archaeologists have to deal with – the physical remains of a field system. Fields, by their very nature, are the products of agricultural practices which are first and foremost intended to disturb the ground. Thus the archaeological evidence they will produce is likely to be far more suspect than the accumulated material from a sealed deposit of, say, a pit, ditch or bank. In particular, only when the physical remains of a field underlie a sealed dated deposit or feature or are overlaid or cut by a datable feature can we be absolutely sure of their period. These criteria are seldom met with and even when they are they are not very helpful in showing the beginnings of the field system under investigation. Thus the strip lynchets excavated at Bishopstone, Wiltshire, covered a Roman ditch (Wood, 1956) and the ridge-and-furrow at Hen Domen passed under the bailey of the eleventh-century castle (Barker and Lawson, 1971), two facts which in themselves are not conclusive evidence of origin. The really early dating necessary to prove the existence of such types of cultivation is still lacking. The work at Gwithian remains almost the sole example of closely dated early medieval fields, but whether the cultivation ridges there are even the result of ploughing remains highly questionable (Fowler and Thomas, 1962).

The limits of archaeological evidence must be stressed. Much is known about prehistoric and Roman fields for they have been identified, planned and excavated. Details such as how they were laid out, what boundaries they had, what types of ploughs were used, what techniques of cultivation were employed on them, what crops were grown on them, and sometimes what settlements they belonged to have become clear in recent years. All this is very important and impressive. But archaeologists do not know and cannot know what, from the historian's point of view, is the most important of all, the overall tenurial arrangements and how the fields were organised by the society which developed and cultivated them. When archaeologists come to open fields, because the historians can tell them at least something about the tenurial background, they can understand the physical remains in terms of divided strips, two or three field systems, and rotations. They can fit the physical remains into the wider system of the agricultural economy that can be shown to have existed by the late medieval period. They must never forget, however, that without the historical evidence ridge-and-furrow, for example, would be totally meaningless beyond the certainty that it was formed by a technique of ploughing. They would never realise the complex pattern of landholding, communal cultivation and social organisation just from the physical remains themsleves. Thus ridge-and-furrow can be of any date and all one needs is a plough of a certain type and to use that plough in a particular way. After all no one would suggest that there was ever an open-field system in Australia, yet there are large areas of good ridge-and-furrow in New South Wales (Twidale, 1972). Of course this does not mean that as archaeologists working in the historic period we cannot use historical evidence. We would be rightly accused of narrow-minded attitudes if we did not use it. But we must not forget that this evidence is historical and not archaeological and therefore is not necessarily explaining what we are dealing with. Clearly there is a link between ridge-and-furrow and the open-field systems in their final complex form, for the historians have shown us this. But whether this relationship is of any significance in understanding the origins of either the ploughing technique or the complex system of agricultural and social organisation that made up the open-field system is by no means certain.

Another difficulty is that fields, again by their very nature, are subject to both long- and short-term changes in purely agricultural practices which may have nothing to do with the overall operation of the open-field system in its wider sense. These changes inevitably result in the destruction of much of the earlier phases by the process of later cultivation. This is, of course, true of all archaeological sites but it is much more apparent with fields than with other remains. Thus the physical manifestations of open fields which archaeologists have to deal with are the result of the pattern imposed by the most recent cultivation, not the first. Between the last and the first may be a thousand years or more of continuous destruction and alteration, year in and year out. This is on a vastly different scale from the apparent 20-year rebuilding cycle of houses in medieval villages or even the gradual denudation of a motte or a hill fort.

Thus archaeologists who look at ridge-and-furrow or strip lynchets, and accept them as medieval, either in form or layout, and see in them the possible origins of an open-field system, do so at their peril. Those who believe that the physical remains that we can see might, even remotely, be connected with the beginnings of the field system, need to be able to put forward very convincing proof.

As a field archaeologist this writer has been, perhaps more than most people, involved in the recording of the complexity of the physical remains of open fields, and in particular ridge-and-furrow and strip lynchets. Yet in the end it would seem that though it is possible to learn much about the history of agriculture from the detailed examination of these remains, it is not easy to understand the origins of the open-field system from them.

There are many interesting features of ridge-and-furrow which seem to reflect certain techniques, events or trends and which are perhaps worth detailing. There is widespread evidence for the overploughing of former headlands to make two, or sometimes three, end-on furlongs into one with the resulting double or treble reversed-S curve. This might be interpreted as a result of the changeover from ox traction to the more powerful horse traction. However this relatively common feature implies considerable changes in the tenurial as well as the agricultural arrangement of the fields at the time it took place. The same reason, that is an alteration of plough traction, is perhaps the answer to the less common occurrence of the partial elimination of reversed-S ridges by the formation of later straight ones. The plentiful evidence for the extension of furlongs into formerly damp or difficult ground, or the opposite, the shortening of furlongs to avoid land which had become waterlogged or difficult to cultivate, might be explained as a result of complex economic or even climatic changes. The process of throwing together narrow ridges to form wide ones or the splitting of wide ridges to form narrow ones, also well attested, may be interpreted as the result of tenurial changes. The overploughing of stabilised land-slips in ridge-and-furrow, very common in some areas, may be a reflection of economic prosperity or of overpopulation. Other features that have been noted, such as the joining up without a break of two, three or four ridges into one, or the relaying of one set of ridges at right angles to another and older set, are more difficult to explain. On a larger scale the evidence for the extension of villages into fields formerly ploughed in ridge-and-furrow, as well as the extension of ridge-and-furrow into areas of abandoned villages or of abandoned parts of existing villages, also reflects the picture now emerging of constant change in the physical layout of open-field systems.

Most medieval archaeologists are familiar with the evidence for Roman sites and even Saxon settlement discovered below ridge-and-furrow. Sites such as West Stow or New Wintles come to mind. However, such total alteration of the landscape also went on at a much later date. There is now growing evidence for the complete overploughing of some deserted medieval villages in the Midlands, and other sites appear to have suffered the same fate. For example at Grafton Regis, Northamptonshire, there is a complex monastic and later manorial site which has been fully excavated (RCHM, forthcoming). The archaeological evidence shows that the abandonment of the site occurred in the late fifteenth century while documentary evidence indicates that it was deserted soon after 1491. Yet air photographs show that the site was later entirely overploughed with ridge-and-furrow, part of one of the later open fields of the area which existed until parliamentary enclosure in 1727. In fact there was almost no trace of the site on the ground before the modern destruction of the ridge-and-furrow and even the hollow-ways passing from the adjacent village through the ridge-and-furrow to the next village totally ignored it.

The evidence from the physical remains of the common fields thus indicates that ridge-and-furrow, and indeed strip lynchets, are the result of many complex changes over the centuries. However, it is probable that most of these obvious results of change and alteration are relatively modern and many are not even medieval at all but are part of the immense economic, social, tenurial and technical changes affecting agriculture in the fifteenth, sixteenth and seventeenth centuries which as yet have hardly been appreciated even by the economic historians.

Nevertheless archaeologists must examine very closely the results of the recent work on open fields by historians and geographers who have shown the complexity of open-field systems, which the slighter work of the field archaeologist whole-heartedly confirms. It is clear from recent literature that, even by the thirteenth century, there was not one type of open-field system but many, adapted to complex geographical, social, tenurial and economic circumstances and differing not only from region to region but from parish to parish. The well-documented late medieval and post-medieval alterations and changes to these systems are even more complex and from the purely archaeological point of view it

would seem that it is very difficult to learn much about the early forms
of open-field systems from the close examination of the physical remains.
Archaeologists are seeing only the final complications of a slowly
evolving system of agriculture which has been subject to many changes.

Is it not possible to go further than this? Can we not see from the
wider, overall patterns of ridge-and-furrow or strip lynchets some clue,
if not to the origins of the open-field system, at least to their early
evolution? Various attempts have been made to see an underlying
picture of development of the open fields from the shape and size of
furlongs. Certainly to some extent minor developments can be suggested.
These include the infilling of awkwardly shaped areas of waste, perhaps
once left out of the main arable system, but later incorporated into it
so producing curiously triangular blocks of ridge-and-furrow. On the
whole, however, these can probably be attributed to relatively late
expansion of arable land. Likewise the marked contrast between areas
of furlongs of regular layout and those which are intermingled and
interlocked is probably better interpreted as an adaptation to the
physical environment than a part of the development of the fields
themselves. In any case even if it is accepted that these fields show
development, they tell us nothing about the origins of the open-field
system itself.

What then can be the archaeological contribution to the problem of
the origin of open fields? There are four important ways in which
archaeological evidence can be used. Perhaps the most important of
these is that of the continuing research into, and the wider
dissemination of knowledge of, prehistoric and Roman fields and
society. One of the major mistakes made by most past, and indeed
many present, non-archaeological workers in the study of open fields is
their failure to appreciate the pre-existing physical and social conditions
in which the medieval open fields developed. In many of the seminal
works by geographers and historians there is still the stated or implied
belief that the medieval open fields originated in a virgin landscape and
were developed by a society which was quite unconnected with all the
social and tenurial, economic and technical encumbrances of their
predecessors in Britain. Yet if prehistoric and Roman archaeology has
achieved anything in the last 20 years it is that it has proved beyond
doubt that from as early as 2000 BC Britain was a well-populated
country, almost totally exploited agriculturally by a sophisticated and
complex society. Certainly by the latter part of the Roman era, most
parts of Britain were densely settled, with fields of some form over
much larger areas than those of the twelfth or thirteenth centuries AD.

These fields were being worked by technically advanced farmers, operating from thousands of settlements, and almost certainly arranged in clearly definable land-units or estates based on a complicated system of tenure.

It was into this landscape and into this society that perhaps a relatively few Saxon people were injected and in the fullness of time developed what can be interpreted as true open fields. Therefore any discussion of the origins of open fields has to take into account the fact that, whenever or however the Saxons developed open fields, these had to be based on a pre-existing system of agriculture and field shapes and did not evolve in an empty countryside devoid of any remains of earlier farming. Whether the existing system or the fields were still in use when the open fields as they are understood finally evolved is a question that cannot be answered at the moment. But the main point is that the open-field system was at least partly based on what was already there.

This brings us to the second way in which archaeologists can use the methods of their discipline, that is to identify and date the earlier pattern of pre-Saxon fields on which the later open fields may have been based. Such work has already started and Peter Fowler and the writer have already published the admittedly slender evidence that was known to us two years ago (Taylor and Fowler, 1978). It was recognised that some of the furlongs in the medieval and later open fields are bounded by ditches, some of which appear to be Roman in origin. This evidence suggests that we may have the beginnings of proof that the basic unit of cultivation in the medieval open fields, that is the furlong or blocks of similarly orientated strips, was based on an older arrangement of fields, probably of Roman origin. This is not to suggest that the ultimate shape of every furlong was that of a Roman field. Indeed there is good evidence that this is not so. All that one can say is that the basic framework of furlongs may have been partly established on earlier fields and that these were subsequently altered by the later developments which have been outlined above.

At the moment the evidence for this hypothesis is weak and needs much more support which it is hoped will be forthcoming. But even if, in the end, the hypothesis is proved, it is still by no means certain that archaeologists have learnt anything about the origins of the open fields. We may finally convince the historians and geographers of the importance of the prehistoric and Roman physical basis of the medieval open fields, but whether we will be able to explain how the latter could or did originate is more doubtful. It is quite possible that the Roman and prehistoric fields were in fact open fields but we have no way of proving it. If open-field agriculture did exist in Roman or even prehistoric times, then all of us, archaeologists, historians and geographers alike, have little hope of ever establishing its origins.

Thus we come to the final way in which archaeologists can aid the understanding of some of the most important later developments of open fields, if not their origin. This will be achieved by collaboration with other scholars and not by archaeologists alone. For example the fully-developed classic common field system, Gray's Midland System, which in a variety of forms came to occupy much of England by the thirteenth century, only occurs in a landscape of nucleated villages. Thus it is likely that this type of open field, i.e. the true common field system, is to be associated with the typical English village. At the moment we certainly do not understand the beginnings of the English village. Recent archaeological and historical work indicates that in many places nucleated villages did not come into being until the ninth, tenth or even the eleventh centuries. In other places they are earlier; elsewhere they are even later. But whatever their origins, the classic common field system probably developed with them and if there are great local and regional differences in the origins of the villages, then the same will apply to their fields.

On the whole, in this paper a somewhat pessimistic attitude has been taken towards the archaeological approach to the origins of the open fields. This is not because the value of archaeological research is suspect but because it is believed that archaeologists should always assess very carefully the information available to them by the nature of their discipline. They must be very clear in their minds to what extent they can infer from the basic evidence of ploughing techniques whole systems of agriculture, tenure and social organisation, and how much they are basing their interpretations of the physical remains on sources made available to them by historians. And this by historians who, to extend the earlier analogy, are travelling in the same direction, not only on adjacent tracks but in a rather better upholstered carriage.

2 THE ORIGINS OF OPEN-FIELD AGRICULTURE – THE ARCHAEOLOGICAL FIELDWORK EVIDENCE

David Hall

Introduction

Open-field cultivation patterns hold a wealth of information about the foundation and development of Saxon and medieval farming communities and their settlements. From fieldwork techniques it is possible to reconstruct these field patterns accurately and completely, so providing a useful source of evidence.

Much has already been written about open fields. Amongst the more notable is the work of Gray (Gray, 1915), the Orwins (Orwin and Orwin, 1938), Bowen in 1961 (Bowen, 1962) and the discussion between Thirsk and Titow (Thirsk, 1964, 1966; Titow, 1965). Beresford (Beresford, 1957) and Hoskins (Hoskins, 1955) first drew attention to the application of observation on the ground linked with documentary work. Recent historical studies have been presented by Baker and Butlin (Baker and Butlin, 1973) and the whole range of open-field terminology is discussed by Adams (Adams, 1976).

The picture of open-field agriculture given by the literature is, nevertheless, not at all clear; descriptions of the basic elements are either entirely absent or inaccurate. What is lacking is a full consideration of the physical evidence linked in detail with relevant documentation: not the study of just a few fields comparing maps and aerial photographs, but the study of a whole parish or hundred, field-by-field on the ground. The subject is ideal for the combined use of the skills of the historian, aerial photographer and field archaeologist. Yet for the most part this co-operation has been lacking. It is quite extraordinary that in 1973 it could be said 'substantial evidence to support Beresford's theory about the pre-enclosure origins of much ridge-and-furrow (1957) has not yet been forthcoming' (Baker and Butlin, 1973, 35). The merest cursory comparison of an open-field map and vertical air photographs shows that Beresford is absolutely correct.

This paper will discuss the archaeological evidence for the origins of open fields, approaching the problem from two different viewpoints. On the one hand showing that medieval fields are very antique and show clear evidence of a major modification before the twelfth century,

22

and on the other, showing that early Saxon settlement patterns are unrelated to ridge-and-furrow systems. First, we must have to hand plans of complete open fields for a large number of parishes.

The Physical Remains of Open-field Agriculture

It is quite clear that ridge-and-furrow can be equated with open-field cultivation from the process of comparing pre-enclosure maps with remains on the ground, or with early air photographs. The examples of Strixton (Hall, 1972) and Wollaston (Hall, 1977), Northants, where the coincidence is exact, have already been published.

Beresford has quoted historical statements such as that of Folkingham stating that *terra elata inter duos sulcos* (1610) and that of John Rouse (1485) who clearly interpreted the ridge-and-furrow surrounding deserted villages as evidence of former open fields (Beresford, 1957).

Lands were ploughed in a clockwise manner more often than anti-clockwise (which was only done in fallow years) so they eventually became ridged up forming the characteristic ridge-and-furrow (Figure 1a). Generally furrows are aligned down the steepest natural gradient to assist drainage, that is unless the terrain is very steep. The movement of the plough caused transference of a small quantity of soil forwards along the direction of motion (as well as from left to right which formed a ridge). At the end of the land the plough was lifted out for turning and the extra quantity of soil was deposited. This simple soil movement can be observed with modern ploughing at the incomplete stage when the outside of the field (left to turn on) is yet to be ploughed; small quantities of soil are dragged from the furrow onto the stubble. Over the years with the repetitive pattern of open-field ploughing these small quantities built up to form substantial heaps at the end of each land; they were called *heads* or *butts,* and are most important archaeologically.

In a permanent-grass modern field containing ridge-and-furrow two features are instantly striking: the undulating ridges themselves and the high banks of plough-moved soil that form the furlong boundaries (the headlands and joints). Both are easy to identify and survey. Modern ploughing of such a field rapidly flattens the ridges, although they can often be seen as soil marks and cropmarks for many years after. The great banks of soil at the furlong boundaries, however, survive as long linear features that can be up to three feet high.

The recognition of these banks for what they are seems to be fairly

**Figure 1: (a) Diagram of the Principal Features of Open Fields (b) Later
Modifications of Open Fields: Shaded Areas were Grassed Down**

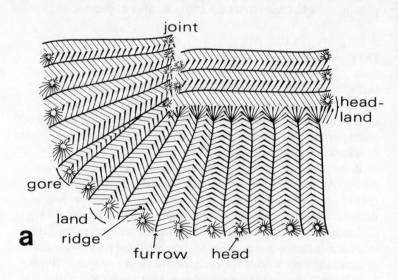

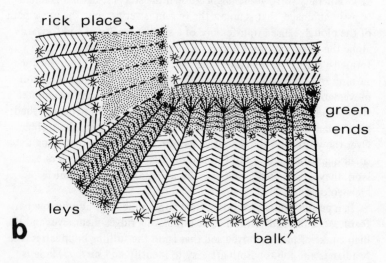

recent. Workers at Maxey in 1960 were rather baffled by this phenomenon. Although it was finally correctly decided that the banks were post-Saxon there was a failure to understand either the mechanism of their formation, or their widespread occurrence and potential as a research tool (RCHM, 1960). In Oxfordshire Pocock observed furlong boundaries in 1963 (Pocock, 1968). Detailed surveys of ploughed medieval field boundaries in Northamptonshire parishes began in 1961 and have been undertaken at an increasing rate so that 90 townships (about 180,000 acres) have now been completed. The techniques for recording furlong patterns have been discussed previously (Hall, 1972); essentially they consist of surveying every field in a parish recording the linear earthworks of the furlong boundaries. The results are supplemented by the use of air photographs; those taken by the RAF in 1946 are particularly useful. It is essential to have the archaeological surveys of an open-field system for detailed study. No surviving contemporary map ever gives the complete picture because of modification and contraction of the ploughed area. In addition there were other changes in the landscape, such as roads becoming established along the furlong boundaries: thus from a map alone the road would be interpreted as a strip of waste or a hollow way, yet really it was an earlier ploughing headland.

The Antiquity of Medieval Fields

As already stated, furlong patterns surveyed on the ground, or visible on air photographs, invariably correspond to open-field estate maps, despite the evidence of Mead's work (Mead, 1954). This comparison has been made in detail for some 17 parishes in various counties (Beds: Barton le Clay, 1798; Cambridgeshire: Burwell, 1806; Northants: Barnack, 1760, Bozeat, 1605, Castor, 1845, Cranford, 1748, Higham Ferriers, 1789, Little Houghton, 1829, Raunds, 1798, Stanwick, 1826, St Martins Without 1773, Strixton 1583, Sutton 1845, Weedon and Weston 1591, Wollaston 1774 and Woodford 1742). It is therefore clear that ridge-and furrow is at least as old as the earliest maps; the finest one so far encountered is that for Strixton of 1583 (Hall, 1972).

Historically ridge-and-furrow patterns can be shown to be medieval where there are detailed terriers with furlong names corresponding to those on open-field maps. This has been done for Wollaston, Northants, where lengthy terriers of 1372 and 1430 list most of the furlong names marked on a map of 1774 (Hall, 1977). All the furlong boundaries shown on this map survive on the ground. Few such detailed studies appear to have been published, although it is easy enough to find names from late-medieval terriers that correspond to furlong names on the open-field maps, implying that most cases would be like that at Wollaston with a long-term continuity.

There is physical evidence showing that some ridge-and-furrow is medieval. The best known being Hen Domen, Montgomery, where ridges are buried by eleventh-century earthworks. At Bentley Grange, Yorkshire, coal-pit spoil heaps of the twelfth century overlie ridge-and-furrow (Beresford and St Joseph, 1958). Various other medieval features cut across or through pieces of ridge-and-furrow unequivocally demonstrating that these areas at least are ancient. At Titchmarsh, Northants, a manorial garden was enlarged to include part of a furlong. On the inner side the lands can still be seen in their low profile thirteenth-century state, and on the outside the remaining half lands were ploughed until 1778 (Hall and Belgion, 1979), forming a new series of heads abutting the park pale. Another example known in the area is again at Wollaston. A document of 1231 describes the garden bounds of one of the two manors stating that it stretched as far as a ditch dug anciently through ploughland. The earthworks of this manor survive intact and the ditch can still be seen curving through low-profile ridges. These lands are twelfth century at the latest, if 'ancient' in the document can be assumed to mean older than 31 years (Hall, 1977).

Thus it is not difficult to show that various pieces of ridge-and-furrow are medieval, and where detailed studies have been made, furlong names on open-field maps can be shown to date back to the fourteenth and fifteenth centuries and in some cases to the twelfth century (see below). The overall physical layout of open fields must therefore have changed but little from this period until enclosure.

During the early post-medieval period the land use and cultivation of open fields was considerably different from earlier times. In particular there was an increased amount of grass without and within the cultivated area. Some of the modifications are shown on Figure 1b. They have been discussed elsewhere (Hall, 1977), and do not directly affect the present discussion.

Contemporary writers were well aware of these large-scale changes of land use. Thus at East Haddon, Northants, it was stated in 1771: 'the cow pasture contains 600 acres of ancient greensoard that has not been plowed up for more than 500 years' (Northampton Record Office, 1L 2122. Referring to the whole of that county in 1712 Morton writes:

Many of the Lordships, and especially the larger ones, have a Common
or uninclosed Pasture for their Cattel in the Outskirts of the Fields.
Most of these have formerly been plowed; but being generally their
worst sort of ground, and at so great Distance from the Towns, the
Manuring and Culture of them were found so inconvenient that
they have been laid down for Greensod.

The furlong patterns in these ancient permanent pasture areas do not
differ from those of the later arable, therefore any changes in furlong
layout must date back well into the medieval period if not before.

Does the furlong pattern give any clue to the origin of open fields?
The final pattern for a given parish is principally dependent upon the
landscape. As strips are generally aligned down the steepest gradient
there will be a large number of small furlongs orientated in all
directions in a region where an undulating landscape is dissected by a
multiplicity of springs and brooks. On plains and terraces, as at Maxey
or Etton, Northants, furlong patterns are simple rectangular systems
(Figure 2). This rectangularity seems to be the objective of an 'ideal'
open-field network.

Beresford has suggested that open-field patterns are arrived at by a
piecemeal reclamation of waste, perhaps taking in a furlong at a time,
getting progressively further away from the village as the population
grew. Many furlong patterns do show evidence of being based on the
present villages, but there are puzzling aspects in the field patterns of
the 90 survey Northamptonshire parishes that are studied. How could
tiny furlongs (marked 'S' on Figures 2 and 3) result from the reclamation
of a chunk of land? Surely they are the result of subsequent changes.
Most curious of all is the way in which some furlong boundaries
gradually cease so that two furlongs merge into one (marked 'A' on
Figures 2 and 3).

There is evidence for two types of modification, one relatively minor
and the other major. At Wollaston, Northants, a small furlong, marked
on an open-field map of 1774 (Northampton Record Office, Wollaston
map, 1774) and surviving as an earthwork boundary feature on the
ground, does not exist as a cropmark. From the air photographs the
strips of the adjacent furlong can be seen to continue through at right
angles to the direction of the small cross-furlong. It is clear therefore
that the minor furlong is a later modification of the system, doubtless
made to satisfy a local gradient (Plate 1). It may be that many such
small furlongs are similar modifications. (The final furlong pattern is
discussed and illustrated in Hall, 1977. The small furlong is No. 87.)

Figure 2: Etton, Northants, Part of the Open-field System

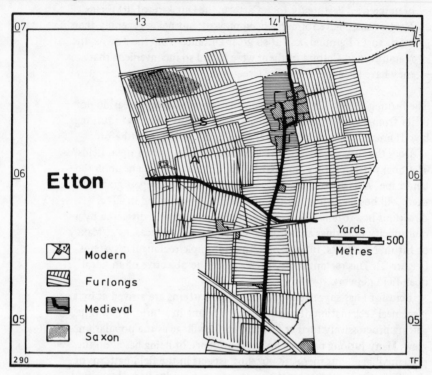

Figure 3: Raunds, Northants, Part of the Open-field System

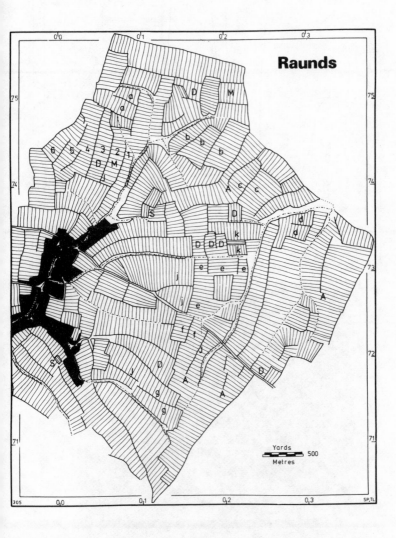

Note: aa, bb, etc. are groups of furlongs with related names such as upper- and lower-; J is a boundary referred to as joint; M is a 'middle' furlong; D is a furlong name that is only a description and therefore presumed to be secondary.

Plate 1: Wollaston, Northants. Part of the open fields, showing the plan in 1774, and an aerial photograph of 1948 showing furrows underlying a furlong boundary

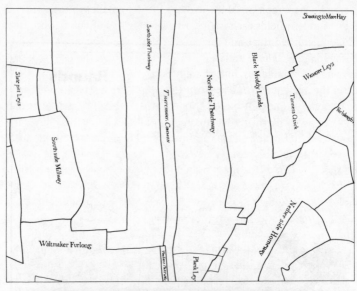

Source: British Crown Copyright/Dept. of Environment photograph.

A major modification recently noticed at Wollaston was that some of the 'present' furlongs have been created by the division of furlongs with much longer strips. Plate 1 compares the 1774 open-field map and an aerial photograph of 1948. The furlongs lie parallel to each other in several modern fields each with a different history of cultivation, from grass undisturbed ridge-and-furrow to fields once-ploughed and many-times ploughed. Those with undisturbed strips show the reversed-S curve at the boundaries, but where modern destructive ploughing has begun the ancient furrows are seen to continue in a straight line under the boundary. The only explanation is that the two furlongs were made out of one that originally had much longer strips by cutting across all the strips and forming a new turning place. The shorter lands gradually developed reversed-S profiles and piled up banks of soil on the new boundary. North of the present road (which follows the ancient track called *Thatchway*) there are three furlongs with well-developed reversed-S lands at the boundaries. The furrows cannot be seen to go straight under the boundaries in many places but there is an irregular sequence of narrow and wide lands showing some similarity in all three furlongs. Although not exactly the same in sequence, there is much more correspondence than can be accidental, so providing further evidence of large-furlong division. Numerous other similar cases can be found in this parish and elsewhere in Northamptonshire (DoE, 1976; Steane, 1974; Belgion, 1979; Hall, 1979). An example at Arnesby, Leics. shows that strips either side of a large boundary 'line up' and have a repetitive sequence of widths (J. Pickering, personal communication).

Once the immutability of furlong boundaries is challenged, significant furlong alignments become clear. At Raunds no less than six furlongs have strips lying on a gentle curve. Although most of the boundaries have developed the usual reversed-S the overall alignment is quite clear: the furlongs have been created by the subdivision of an enormous one with curved strips 1,100 yards long. On air photographs this curvature is quite clear and the boundary between furlongs 1 and 2 (Figure 3) is quite invisible, although it is plain enough on the ground. The problem of furlong boundaries suddenly finishing now resolves itself: they are the later rather arbitrary divisions. Figure 4 shows the conjectural original open-field layout of part of Raunds, Northants. In many cases there is evidence from aerial photographs (or from the furlong names, see below) that the pattern of long furlongs shown is authentic. Such long strips up to 1,000 yards length are not without precedent. At Wharram Percy and Burdale, Yorks. there are large furlongs with lands up to 600 yards long running from dale to dale over the chalk plateau; these have never been subdivided. In the Wisbech region of Cambridgeshire, particularly at Parson Drove and Elm, there are unridged ditched-strips up to 1,300 yards long (Hall, 1978). In both these regions the soils are very light and it may be that it was convenient to work large lands, whereas in the heavier soils of the Midlands shorter lands were necessary so that each one could be ploughed in a day and there were more resting places for draught beasts when they turned. On the other hand long strips are known at Holderness, Yorks., where the soil is heavy (see below, p. 185).

Furlong names give some supporting evidence of division. So often they come in pairs, such as *Upper* and *Nether Whines* for example, or more significantly the furlongs *Above* and *Below Holme Way Joint*. Joint is the furlong boundary, but it may reflect the memory that the large furlong had been split along this point. At Raunds, Northants, there are many furlongs with the appellation 'middle', as well as 'upper' and 'lower'. Some of the furlong boundaries are referred to as first, second or even third joint. All this is evidence of a later subdivision and re-planning (Northampton Record Office, ML 124, 1739).

At what date was the subdivision of furlongs made? There could be dangers in relating just a few medieval furlong names to a late map and assuming complete correspondence. In 1372, at Wollaston, there was a furlong above Thatcheweis and a furlong below it, but how can it be decided whether these names refer to large early furlongs or the smaller later ones? In this case it can be resolved from a long detailed terrier of 1430 (Hall, 1977). In the Nether Field there is, in 1774, only one furlong name containing the element 'black' — *Black Muddy Lands*. The 1430 terrier lists just one also, *Blackmanlond*, which is therefore likely to be the same. There are 13 separate entries for this furlong in the terrier, one of which gives the significant variant, *Black Middlelond*. As this furlong is the middle one of three furlongs formed by subdivision (Plate 1) we can see that the identification is sound, and that by 1430 the significance of the word *middle* had already been forgotten. The *'black'* element comes from the dark soil caused by the presence of a buried Romano-British site there.

There is other evidence that the subdivision was much earlier than the fifteenth century. A feature noted by fieldwork is that some small later furlongs have their common boundary ploughed over, linking the two strips approximately in line, into one. There is usually a kink at the junction and, of course, the hump of the old headland boundary still remains. In other words there is a reversal of the division of long furlongs.

Figure 4: Conjectural Original Layout of Raunds Open Field

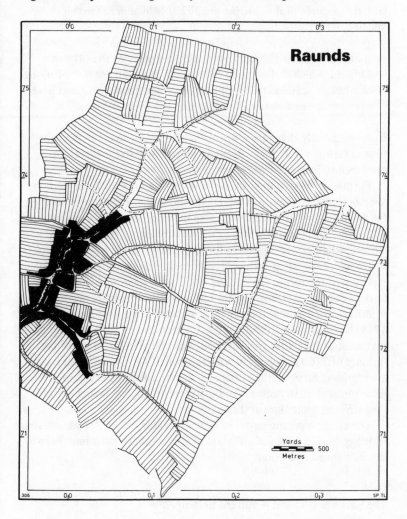

It had been assumed that this was a post-medieval feature, but there is an early example of it occurring in 1252 at Salden in the parish of Mursley, Bucks:

> Item below Little Pusshulle (furlong), one piece of the tithe of Luffield, which furlong used to be on its own (but) now the furlong of Abswale is added to it by ploughing, thus the two furlongs have been made one that used to be divided (Elvey 1975).

It seems unlikely that such a process would be occurring at the same time as furlongs were being divided up, and the latter is therefore presumed to have occurred much earlier.

Further evidence comes from the widespread use of the name *Long furlong*. Inspection of open-field maps shows that the name is reserved for furlongs with strips of up to twice the usual length, i.e. approaching 400 yards. This use of the word long would have no relevance in a landscape laid out with massive furlongs of the type shown in Figure 4. Therefore the 'great furlong' layout must necessarily antedate any early use of the name *long furlong* in the sense described above. A terrier of about 1160 for Cranford St John, Northants (Stenton, 1930), mentions nine furlongs that are identifiable on an open-field map of 1748 (Northampton Record Office, Cranford Map 1748). It is clear that most of the parish (probably all) was under cultivation and the furlongs of 1160 are identical with those of 1748. One of them is a *long furlong*. So again the evidence points to a furlong subdivision prior to the twelfth century, and therefore an original laying out of long strips at some time in the Saxon period.

The most surprising aspect of all is the laying out of the original long furlongs. Such a large-scale division of the landscape must have been a single, planned operation.

The Saxon Settlement – Pattern Evidence

Recent archaeological evidence for early Saxon settlement in Northamptonshire weighs against the early introduction of strips. Many Roman villa sites yield a few Saxon sherds, suggesting an initial continuity, and in addition there are many small Saxon sites, away from present villages, settled *de novo*. The distribution of such settlements is mainly restricted to light soils of river gravels and Northampton sand and ironstone, and is strikingly similar to the Bronze

Age settlement pattern (Hall and Martin, forthcoming, a). This
alone implies settlements with primitive agricultural techniques,
incapable of utilising the heavy clays that had been ploughed by Iron
Age and Roman peoples, and unlikely to be founding the village pattern
so familiar by Domesday with settlement on clay land. A further
objection to strips being associated with these early sites comes from
the furlong patterns themselves. If there were furlongs (or as yet
undivided plots corresponding to furlongs) associated with these
settlements which eventually were incorporated in the open-field
pattern for the present-day villages, then they should be clearly visible;
but they never are. There is much Saxon settlement at Brixworth (Hall
and Martin, forthcoming, a), and in the Welland Valley north of
Peterborough (e.g. Figure 2), but the furlong patterns do not reflect it.
Nor do they now at Doddington (Foard, 1978) and elsewhere. The
early Saxon sites, along with all those of earlier periods, are ploughed
over by furlongs. Therefore it is unlikely that the communities living at
these settlements were directly responsible for the open fields.

The continuity of field systems must be mentioned. If the early
Saxons were not responsible for furlongs it is perhaps specious to look
for continuity between 'celtic fields' and medieval fields on a large
scale. In vain one looks for many extensive systems of Roman or
prehistoric fields that could convincingly be converted to a recognisable
furlong pattern (Bowen and Fowler, 1978). One of the best known
large Roman 'field systems' north of March, Cambridgeshire, is no more
than an extensive series of cattle-raising paddocks associated with a fen
saltern industry. It can hardly be said to be typical and is most 'un-
furlong-like' (Taylor, 1975). Detailed fieldwork has been carried out in
the adjacent parish of Elm. The complex cropmarks continue near
formerly tidal watercourses, but farther north there are Romano-
British settlements without any cropmarks (which should be clearly
visible on the light silt soil if there were any buried ditches). The
implication is, therefore, that normally the Romans farmed some kind
of flat-ploughed unditched 'open' field system (Hall, 1978).

In Northamptonshire, therefore, there are two problems. On the one
hand the small early Saxon sites were deserted to form the present
nucleated villages, and on the other the landscape was divided up on a
massive scale into strips. Did these two operations occur together, each
necessitated by the other? The precise date of such a change in settlement
patterns is unknown, but the early Saxon sites produce no Saxo-
Norman wares and so were presumably deserted before that pottery
was introduced in the ninth century. Generally the pottery from these

sites is amorphous and undatable, but most of it has close parallels with the Middle Saxon material from Maxey (Addyman, 1964) and few relationships with early Saxon material, e.g. that from Orton Waterville, Cambs (D. Mackreth, personal communication). It seems unlikely that such a remarkable re-planning of the landscape would occur before there was a considerable settled population exerting pressure on land, so that a late eighth-century date is suggested for both the desertion of the Saxon sites and the first formation of strip fields.

If the present villages were formed by nucleation of several smaller settlements before the ninth century, then it follows that handmade Saxon pottery should be found at their present sites. Careful monitoring of building works in village centres has shown that nine out of a sequence of fifteen villages on the south side of the Nene Valley, Northants, do indeed produce Middle Saxon sherds.

Various excavations have produced results that also agree with the dating; for instance at West Stow, Suffolk, a Middle Saxon settlement was found underneath ridge-and-furrow which itself had been buried by a mantle of blown sand in the thirteenth century (West, 1969).

The eighth-century date fits in with what little clear historical evidence there is. The information from late Saxon charters strongly suggests that the 'medieval' strip system was already in physical operation by the tenth century (Finberg, 1972; Gelling, 1976, a). At Newnham, Northants, a boundary charter of 1021-3 describes furrows and headlands that correspond to right-angled bends in the present-day parish boundary (Gover, Mawer and Stenton, 1933). An open-field map of 1764 shows that the kinks are caused by the presence of furlongs (Northampton Record Office, Newnham open-field map 1764).

Conclusion

The origins of open fields have been discussed for many years. On the one hand there was the idea accepted by Seebohm and the Orwins that the Saxons introduced open fields on their arrival in the fifth century, and on the other hand more recently Joan Thirsk has argued that early Saxon communities held their land in severalty and later, because of population pressures (and partible inheritance) holdings were divided up into strips and eventually farmed communally (Thirsk, 1964). More detailed considerations have been given by Campbell and Dodgshon (pp. 112-44).

The archaeological evidence given above falls between two of the contending views on the formation of open fields. They do not seem to have been laid out by the early Saxons, but rather later, nor do they seem to be subdivided early private holdings, but a strip system *ab initio*.

The reasons for such an unrecorded radical change of settlements and fields are a matter of conjecture. Postan attributes strip fields to the widespread use of the heavy plough and the requirements of collective landholdings (Postan, 1972). The Church, too, may have been instrumental and other possible causes are discussed in this volume.

There are various lines of research to establish further the validity and geographical extent of the long-furlong phenomenon. Detailed ground surveys are essential to establish complete (small) furlong maps; they cannot be compiled from aerial surveys alone, nor from extant comtemporary open-field plans. Cropmark vertical aerial photography is required to see if furrows go under the later furlong boundaries and so establish which boundaries belong to the original layout.

A variety of information may come from the corrected early furlong pattern. Depending on the date of furlong division it may be possible to see the centre part of a parish with long furlongs and the outer parts laid out with normal small ones representing the type of planning before and after division date (see Campbell, below Ch. 5).

The problem of field-system continuity from Roman to medieval furlongs now has to be reconsidered afresh. On the whole, continuity seems unlikely except perhaps where a long furlong boundary was aligned on some remnant of an older earthwork, etc. A re-planning of the landscape would account for the general absence of Celtic elements in medieval field names. For a later analogy the loss of medieval furlong names at parliamentary enclosure is well known. Normally modern field names preserve few of the open-field names.

A few furlong boundaries with ditches under them have been reported (Bowen and Fowler, 1978, 159), and another has recently been observed at Little Houghton. These may be ancient features or maybe long furlongs that were initially marked out with ditches. A large landscape study is required in areas being totally quarried away in order to provide the opportunity to study all the furlong boundary sections.

Major landscape changes, therefore, seem to have taken place during the Middle Saxon period: settlement on heavier soils, the laying out of long furlongs and their subsequent subdivision. Also there was the abandonment of small scattered settlements with concomitant nucleation into villages. These two processes were probably interlinked, being a response to increasing population, increasing demands for royal and ecclesiastical taxation, and increasing use of the heavy plough.

It is likely that the laying out of the open fields was substantially completed during the eighth and ninth centuries, continued pressures leading to their subdivision before the Norman Conquest.

3 OPEN-FIELD AGRICULTURE — THE EVIDENCE FROM THE PRE-CONQUEST CHARTERS OF THE WEST MIDLANDS

Della Hooke

An understanding of early farming systems is not readily obtained by archaeological methods and, even if it were, such data is sadly lacking for large areas of the West Midlands. My work extends over the former counties of Worcestershire, Warwickshire and Gloucestershire, and much of this area would remain a blank upon the map of Anglo-Saxon England were it not for the evidence contained in place-names and charters. While place-names offer invaluable information about the general characteristics of the early English countryside it is the charters which provide the finer detail. They cast one immediately into the midst of the rural community and provide some of the earliest detailed documentary evidence of field systems and farming practices.

Much of this evidence is contained within the body of the charters but even more is contained within associated boundary clauses, which make their way around the perimeters of estates noting prominent features and naming them as boundary landmarks. While questions of land donation, confirmation and inheritance in charters may be seen more rightly as the concern of the aristocracy, whether lay or ecclesiastical, the boundary clauses, even if composed by clerics, inevitably reflect the world of lesser men. The language of the clauses may be stilted and stereotyped but one is instantly confronted by the immediate world of the early medieval peasant and his daily concern with the fields which supplied both his food and his contribution to lord and church and the woodlands in which he pastured his swine and gathered his fuel. We hear of the paths traversed by the plough oxen and the trees which gave shelter to his sheep. In reference to stream and hill we are made aware of the close relationship then still existing between man and his environment and may trace the relationship of township communities to the surrounding topography.

It is perhaps the apparent fragmentary and parochial nature of the evidence that has deterred so many from examining the clauses further, but information that can be gleaned from them can be used systematically to cast light on much wider problems than the reconstruction of the local landscape. One way in which the data has been used has been to map selected features and to compare their distribution with equivalent distributions of place-name terms. The West Midland area is fortunate in possessing a large number of charters with associated boundary clauses (Figure 5) and comparisons may be made which take into full account any bias attributable to the influence of land ownership or to the date of the documents. The Church of Worcester, for instance, held estates widely scattered throughout the area and preserved many documents concerned with the leasing of individual estates, yet the distribution patterns to be discussed remain clear within these lands alone. Within the region it soon becomes apparent that agricultural features are recorded most frequently in particular areas and I believe this to be of fundamental importance in understanding the pattern of Anglo-Saxon agriculture.

Interpretation of OE Terms

Before proceeding further, however, the terms must be examined with some care, for the interpretation of Old English terminology is by no means as straightforward as it may at first seem. Meanings are known to have changed over the generations and it is only possible to interpret the OE meaning after a thorough investigation of the available evidence. A number of words are frequently understood to be related to agricultural activity, among them *aecer, furlang, furh, gāra, hēafod* and *hēafod-land, hlinc* and *land,* many of them familiar terms to those used to examining medieval field systems. In a later period the *aecer* and *land* was often a strip-holding in the open field, and a *furlang* a group of such strips, while the plough oxen were turned on the *hēafod-land* or 'headland'. A *hlinc* or 'lynchet' was formed between strips on steeply sloping ground and the gore, from OE *gāra,* was a triangular-shaped piece of land left at the corners of irregularly-shaped fields (Adams, 1976, 80-91). One must beware, however, of accepting the charter evidence of *aecer, furlang,* etc. at face value or as proof of a fully evolved open-field system.

The original meaning of *gāra* was 'a wedge-shaped piece of land' after the OE *gār* 'a dart or spear' (Toller, 1921, 283), and its use at an early period need not have had any agricultural connotation. King Alfred in fact referred to a corner of Spain as a gore (Sweet, 1883). An example from the West Midland charters concerns a plot of land lying beside the Ryknield Street at the extreme southern end of Willersey parish (Figure 6). The gore was a plot of land near the junction of two major routeways

Figure 5: West Midland Areas with Boundary Clauses

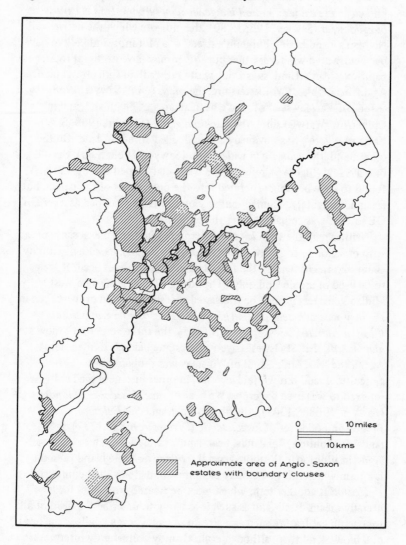

0   10 miles

0   10 kms

Approximate area of Anglo - Saxon
estates with boundary clauses

where they reach the top of the Oolitic limestone escarpment, a plot known as the *ealdegare quod indigenae nanesmonnesland vocant secus buggildestret* (Sawyer, 1968, S.80) 'the old gore which the natives call No Man's Land beside Burghild's street' — a plot most unlikely to have been cultivated whatever other purpose it may have been put to, its name 'No Man's Land' suggesting waste rather than agricultural land. Again, in north-east Worcestershire, the *haeth garan* (Sawyer, 1968, S.786) of Yardley seems to have been an area of heathland, and in Oldberrow, Warwickshire, the *reschitan garan* (Sawyer, 1968, S.79) 'the rushy gore', was obviously a waterlogged corner of land. On the other hand, the *ealdgare in willerseiam* (Sawyer, 1968, S.80) 'the old gore of Willersey' (Figure 6), referred to a triangular-shaped plot of land later found to have been incorporated into the open fields of the parish, but it cannot be safely assumed that any plot of land described as a *gāra* in OE sources was actually 'under the plough'.

Neither may the term *land* be interpreted in its medieval sense of 'a strip or selion in the open field' (Adams, 1976, 90), in spite of Grundy's claim that in charters 'in the few cases in which it does occur it seems to be used of arable land only' (Grundy, 1922, 61). In some West Midland charters it must be translated as 'estate, landed property', as in *tha thry aeceras maede on afan hamme the sče. oswold geaf bercstane into tham lande* (Sawyer, 1968, S.1405) 'the three acres of meadow at *afanhamme* that St Oswald gave to Bercstane belongs to the estate'. Again, the 'No Man's Land' of Willersey was probably not agricultural land but waste. However, in other instances the land referred to was used for crops. We hear of *thaes bisceopes at londes* (Sawyer, 1968, S.1347), 'the bishop's oat-lands' at *Caldinccotan*, Bredon and 60 acres of *hwaetelande* (Sawyer, 1968, S.1332) or 'wheat-lands' at Stoulton, flax-lands, bean-lands, barley-lands and even woad-lands, in addition to the more general *erthland* or 'ploughland', and this is a term which must be closely examined in its individual context.

Hēafod is another term which must be treated with care, for it literally means 'head' and as such is as likely to describe the head of a topographical feature such as a stream or valley, even a hill, as the head of a ploughland strip, although again Grundy consistently interprets it in the latter way. This is certainly misleading for in the West Midlands this term often refers to topographical features, either as a hill summit, as in Evenlode, or as the head of a combe, as in Withington and Olveston, or as the source of a stream, as in Broadwas and Pensax, and this topographical usage seems to be the commonest one in this area. There is also a very literal use of the term to refer to an animal's head.

Figure 6: The Boundaries of Broadway and Willersey

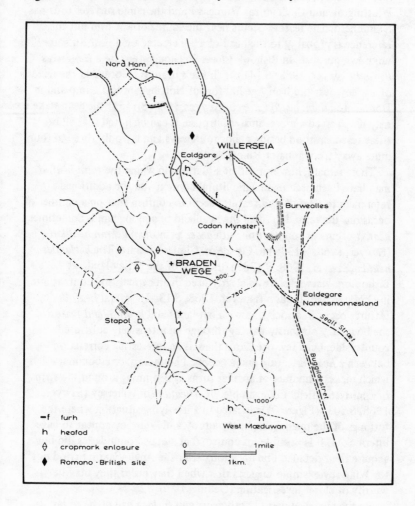

Swines heafde (Sawyer, 1968, S.1329) lies on the boundary of Whittington and Cudley near Worcester and the name survives today as Swinesherd. The feature occurs near the source of a stream but the reference is probably to the head of a pig erected on a post for some unknown purpose. In Bishop's Cleeve in Gloucestershire a *Hengestes heafod* (Sawyer, 1968, S.141) or 'horse's head' is recorded at the head of a valley near the Iron Age hill fort of Nottingham Hill Camp and in Donnington *tham heafod stocce* (Sawyer, 1968, S.115), 'the head stake', may have served a more sinister purpose. It can be noted that all of these features stood beside major routeways and lay only three to four miles away from a major Roman centre.

There remain, however, other situations in which the term *hēafod* may have been used in an agricultural context and the additional references to *hēafod-land* features seem to confirm that on a number of occasions the term described the headland of a ploughland. Sometimes *hēafod* occurs in conjunction with *aecer* as in *tham aecran hēafdan* (Sawyer, 1968, S.1335) in Cutsdean, Gloucestershire. The *berlandes heâfda* (Sawyer, 1968, S.1383) 'the head of the barley-land', at Bishopton, Warwickshire, is clearly used in an agricultural context and in *Caldinccotan*, Bredon (Sawyer, 1968, S.1347), several *hēafod* features occur in association with oat-lands and *hēafod-land* features, the boundary obviously passing through an intensively cultivated countryside. On a few occasions there is an interesting correlation between a *hēafod* feature and a 'stepped boundary' or a boundary which makes a number of angular turns as if running around the strips of a cultivated field. For example, *than heafdan* of Willersey (Sawyer, 1968, S.203) (Figure 6) is located in a low-lying situation where the furlongs of the open fields of Broadway and Willersey appear to have interlocked. It is possible, of course, that stepped boundaries did not acquire their detailed line until long after the Anglo-Saxon period but the Willersey example suggests that when they occur they may be worthy of closer investigation (Gelling, 1976, 626-7).

By the medieval period both *aecer* and *furlang* had come to be regarded not only as measures of land but as measures of land in an open field. The *aecer*, or more specifically, the 'strip-acre' was often equated with the strip or selion and the *furlang* was a parcel of such strips (Seebohm, 1883). Ideally the acre-strip should be 22 yards broad and 220 yards long, the length equated with the furlong as a linear measure, but in practice this varied regionally with the accepted length of the rod used by the ploughman to drive his oxen (Adams, 1976, 2). Aelfric makes it quite clear that a team of oxen in the Anglo-Saxon

period were expected to plough an entire acre in the space of a day
(Thorpe, 1846). The question is whether *aecer* in the charters always
referred to ploughland and the evidence of this region suggests that it
did not, but that it was often being used as a measure of land. In its
latter sense *thritigan aeceran* (Sawyer, 1968, S.1590, S.823, S.80, S.1599,
S.1664), 'thirty acres' of land on the eastern boundary of Hampton
near Evesham may be identified most strikingly with an area today
measuring 550 × 240 yards. This would approximate very closely to a
parcel of thirty-acre strips lying side by side in a north-south direction
if the *aecer* strips were indeed roughly 220 by 22 yards. But while we
undoubtedly have acres of *erthland* or 'ploughland' and, more
specifically, acres of bean-land, wheat-land and flax-land, plus a
reference to *XX aecerum gesawenes cornes* (Sawyer, 1968, S.1423) '20
acres of sown corn', at Norton, Worcestershire, we also have numerous
incidences of acres of meadowland such as the *XII aeceras . . . fulgodes
maedlandes* (Sawyer, 1968, S.1280) '12 acres of very good meadowland'
attached to an estate at Worcester. These occur especially as additional
parcels of land leased with estates, and meadowland, while it may have
been doled out in measured strips, is unlikely to have been cultivated.
The *furlang* in West Midland charters appears less as a linear measure
than as a parcel of land (Gelling, 1976, 627) and it does seem related to
agricultural land. In addition to the furlong of ploughland held by the
monastic community at Pendock in Worcestershire (Sawyer, 1968,
S.1314) a charter of Adlestrop in Gloucestershire refers to *Rahulfes
furlung quae est in campo de Eunelode* (Sawyer, 1968, S.1548),
'Rahulf's furlong which is in the field of Evenlode'. This seems to be a
term found in this area only in an agricultural context.

One other term seems to be safely indicative of cultivated ground.
This is *furh*, 'a furrow'. This may be the most important of all the
terms relating to agriculture for it appears that a furrow was used to
demarcate a boundary where cultivated land reached it on both sides.
In the nomenclature of the open field a furrow was 'a narrow trench
made in the earth with a plough . . . sometimes used to separate strips'
(Adams, 1976, 87) — it was essentially a man-made feature. When
literary evidence is examined the term *furh* does appear to have been
used regularly in a farming context (Hooke, 1978-9, 3-23) and
consistently to describe the furrow opened up by the plough. In the
charters the furrows are described as 'deep, hollow' or 'scooped-out'
and an association with drainage, either deliberate or accidental is
suggested by the *waeterfurh* (Sawyer, 1968, S.1553) 'water furrow' of
Maugersbury, Gloucestershire, although elsewhere dry furrows have

been noted. But the furrows do not always follow lines of natural drainage and may follow lines that are obviously artificial; the bounds seem to have followed furrows rather than crossed them and demarcation would seem to be a primary function, but demarcation connected with agricultural land. In *Caldinccotan,* Bredon, the boundary ran *andlong fyrh anbutan th' hēafod lond* (Sawyer, 1968, S.1347), 'along the furrow around the headland', a furrow bounded the bishop's oatlands in the same area, and in Evenlode Rahulf's furlong, too, is shown by a second charter to have been bounded by a furrow (Sawyer, 1968, S.1325). Many of the furrows noted in the charters lie alongside later medieval open-field systems. In Broadway a large open field called Shear Field extended to the parish boundary to meet the open field of the neighbouring parish of Willersey at the line marked by a *fura* (Sawyer, 1968, S.786) and in the eighteenth century the inhabitants were reminded that they were legally obliged to maintain the hedges and ditches which bounded Shear Field.[1] Today a deep ditch lying between double banks can still be seen on the upper slopes of the scarp face although it becomes less distinct at lower levels. Similarly the *furena* (Sawyer, 1968, S.1599) of the Evesham charter marked the southern boundary of Great Collin Field in Broadway and the fields of Willersey appear to have extended over the area immediately to the north. It is difficult to know the percentage of land which lay under the plough in the Anglo-Saxon period but this problem is connected with the second part of this discussion, namely the evidence of distribution patterns.

Distribution Patterns

When those features which are most likely to have been connected with agriculture are plotted it becomes evident that they are concentrated in particular areas. They occur most frequently in north-east Gloucestershire and south-east Worcestershire and there are hints that this zone should be extended to cover the mid Avon valley of Warwickshire (Figures 7 and 8). Such a distribution is not to be explained by the influence of land ownership or the survival of charter evidence, or even by the date of the documents themselves, all factors which have been thoroughly assessed.[2] One of these areas will be examined in greater detail.

In north-east Gloucestershire on the north-eastern edge of the Cotswolds settlements had been established by the Anglo-Saxon period in the valleys of the Evenlode and Dikler, the latter a tributary of the

Figure 7: The Distribution of (i) Agricultural Features

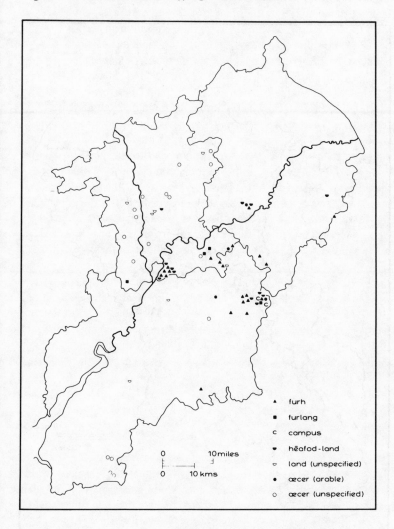

Figure 8: The Distribution of (ii) the *Furh* Term

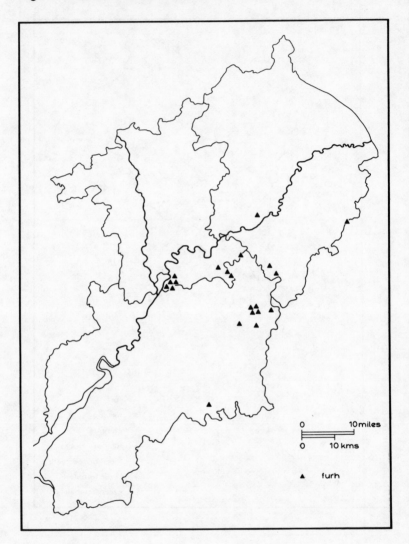

Figure 9: The North-east Cotswolds

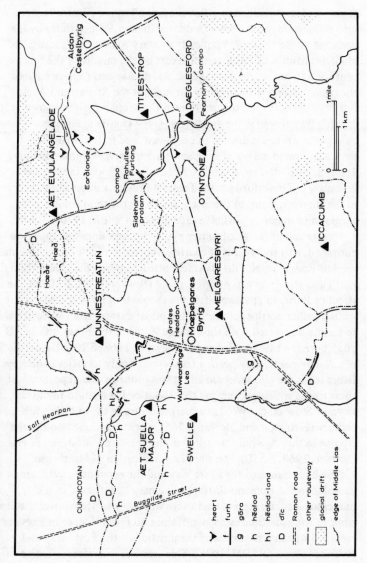

Aldan Cestelbyrig

TITLESTROP

DÆGLESFORD

Fearham campo

AET EUULANGELADE

Eorðlande campo

Rohultes Furlong

Sideham pratam

OTINTONE

ICCACUMB

Hæðe Hoæð

DUNNESTREATUN

Grates Healdon

Mæpelgares Byrig

MEILGARESBYRI'

Salt Hearpan

Wulfweardinge Lea

FOSS

AET SUELLA MAJOR

SWELLE

CUNDICOTAN

Buggilde Strǣt

1 mile
1 km

heort — f
furh — f
gāra — g
hæfod — h
hæfod-land — hl
dīc — D
Roman road
other routeway
glacial drift
edge of Middle Lias

River Windrush which in turn flows into the Thames (Figure 9). Access into this area may have been facilitated in later Roman times by the presence of the Fosse Way, an early military road running across the Midlands from south-west to north-east by the middle of the first century AD, but numerous other routes also appear to have entered the region, many of them in use throughout the Roman and Anglo-Saxon periods. Part of the pre-Roman Cotswold ridgeway known as the Jurassic Way traverses the area, crossing the Oolitic limestone escarpment from south-west to north-east, and in the Anglo-Saxon period it formed part of the *regiam stratam de Norhamtun* (Sawyer, 1968, S.1548) 'the royal road to Northampton' along the northern boundary of Daylesford parish. These and additional routes are recorded in the charters, many of them referred to by the term *strǣet*, suggesting a major or possibly made-up road (Smith, 1970). The concentration of recorded routeways in this area does not appear to be entirely due to the wealth of charter evidence available. An additional road of Roman origin, the Ryknield Street, entered the region from the north and was known as *buggilde strǣet* (Sawyer, 1968, S.1026) 'the street of Burghild', perhaps after the daughter of an Anglo-Saxon king, while another north-south route, the major saltway from Worcestershire to Lechlade, crossed both the Jurassic Way and the Fosse, and in AD 1055 bore the title of *salt hearpan* (Sawyer, 1968, S.1026), a corruption of *salt here-paeth, here-paeth* a term applied only to major routeways. Other roads are also recorded and it was probably the importance of these routes which encouraged the Abbot of Evesham to found the town of Stow-on-the-Wold as a market centre in the mid-eleventh century (Elrington and Morgan, 1965), a centre founded on the site of a possible Iron Age hill fort called in the charters *Maethelgares byrig* (Sawyer, 1968, S.550), literally 'the fortification of Maethelgar'.

The importance of the Fosse Way and the economic advantages it provided in the Romano-British period may be reflected in the numerous settlements and enclosures which lay scattered over the dip slope of the limestone escarpment, since no fewer than eight sites are recorded within a few miles of the junction of the Fosse Way and Ryknield Street (RCHM, 1976). This area was, therefore, a region in which clearance and development were already well established by the early Anglo-Saxon period. Undoubtedly much woodland remained or had regrown. On the hills east of the Evenlode there was sufficient woodland in later Anglo-Saxon times to provide shelter for wild deer, giving rise to a hart's spring, stream, hill, valley and bridge in Evenlode and Adlestrop (Sawyer, 1968, S.1325, S.1238, S.1548, S.550).

Westwards woodland on the scarp and crest of the main Oolitic
limestone escarpment is shown by charter and place-name evidence to
have harboured not only the deer but the wolf and wild-cat, and it
seems possible that the name 'Cotswold' itself may derive from the OE
wald, meaning 'woodland' (Hooke, 1978).

The agricultural land recorded in the charters appears to have been
extensive. It also seems to have been related closely to the geology and
topography, with arable lands located upon gently sloping terrain where
the soils were amenable. Agricultural features in this region include a
number of *hēafod-land* landmarks, and several references to *erthland*
and *furlang.* Some of these features are stated to be part of a *campus*
or 'open field'. In Daylesford the *campo de Deilesfort* (Sawyer, 1968,
S.1548) extended to the north of the present village over a gently
sloping hillside of Middle and Upper Lias and Oolitic sandy limestones,
soils much more easily cultivated than the heavy Lower Lias clays
which cover much of this region, and in Evenlode *Rahulfes furlung
quae est in campo de Eunelode* (Sawyer, 1968, S.1548) occupied a
sloping hillside of Lower Lias overlooking the River Evenlode where the
slope helped to alleviate the otherwise heavy nature of this soil type.
Here the *pratum quod vocatur Sidenham,* 'the meadow which is called
Sidenham', represented the actual riverside meadows. Ploughland seems
to have extended over most of the eastern bank of the river, over a mile
south of Evenlode village itself, and the *threo aeceras earth landes*
(Sawyer, 1968, S.1325), 'three acres of ploughland', show that
agriculture had pushed already into the higher woodlands. It is in this
region that furrows lay alongside furlongs and probably divided the
arable lands of neighbouring communities. Similar development
characterised the valley of the Dikler in Donnington and Upper Swell
where furrows suggest that the ploughlands of these two townships
already met near the latter village, giving, incidentally, the
characteristic stepped boundary of adjoining headlands at the spot
where a *hēafod londe* (Sawyer, 1968, S.115), 'a headland', and a
bradan furh (Sawyer, 1968, S.1026) or 'broad furrow' are recorded in
charter clauses. Here too, the agricultural land already extended some
considerable distance west of Donnington village, climbing up on to the
lower slopes of the limestone escarpment, and above the village of
Cutsdean, where the *aecran heafdan/aecra heâfdan* (Sawyer, 1968,
S.1335, S.1353), 'acre headland', lay on the lower slopes of Cutsdean
Hill, recorded as the boundary which descended to the River Windrush.
There is less information about the nature of the landscape on the higher
parts of the escarpment. In the east of Donnington parish a broad belt of

gravelly drift obviously supported heathland for this is recorded on numerous occasions in the boundary clauses. Westwards, on the higher parts of the Oolitic limestone escarpment itself, agriculture appears to have given way to pastureland and woodland and there are signs that the former was already gaining precedence over the latter under the impetus of wool production. But if agriculture was well in evidence in the Anglo-Saxon period it seems not to have been an innovation and this region appears to have been an area of relatively intensive land use well before Anglo-Saxon times.

This region of north-east Gloucestershire is but one of the areas in which a high incidence of agricultural terms has been noted, and it is not the only one in which the agricultural potential appears already to have been recognised at an earlier period. On the southern bank of the Avon in the Vale of Evesham the Abbey of Evesham was to accumulate extensive estates which may have had earlier antecedents (Cox, 1975). Here many of the agricultural features are recorded in the rather doubtful charters of Evesham Abbey but a sufficient number of authentic clauses exist to show that they were founded upon genuine surveys. The agricultural development seems to have been concentrated along the foot of the Cotswold scarp where intermingled soils of limestone and Lias clays ensured both fertility and good drainage. Here, on the eastern boundary of Broadway, a series of furrows separated the fields of Broadway from those of Willersey, fields lying upon the lower part of the scarp face. Similar lands at the foot of the Oolitic limestone outlier of Bredon Hill are suggested by a clause of *Caldinccotan* (Sawyer, 1968, S.1347) in Bredon in which agricultural features abound as landmarks. In the Vale of Evesham, however, agricultural activity was not restricted to the foothills of the escarpment, for agricultural terms are also recorded on the Lower Lias clays of the vale itself, both on the boundary of Poden in Church Honeybourne, as *thâra aecera aefdum* (Sawyer, 1968, S.1591a), 'the head of the acre', or as *thâm furlunge* (Sawyer, 1968, S.1664), 'the furlong' of Bengeworth where agricultural land seems to have been associated with a settlement *Pottingtûn poticot* (Sawyer, 1968, S.1664, S.80) which has failed to survive. Once again, this is an area of known early development. Romano-British farmsteads are no less numerous than in the vicinity of the Gloucestershire Fosse Way and a probable Roman road runs across from Hinton-on-the-Green towards the Ryknield Street and probably beyond it in a north-easterly direction, towards the mid-Avon valley of Warwickshire. The association between Romano-British sites and cultivated land is particularly apparent in this area. The furrows appear to have run alongside the

cultivated fields of the Romano-British farms and a furlong in
Bengeworth seems to have been associated with another Romano-
British site (Turner, 1973) (Figure 10).

The way in which these areas differ from others is in the high
incidence of agricultural features noted, for the pattern does not occur
in all areas, however rich the amount of charter evidence available. Over
most of Worcestershire, for instance, the clauses reveal the presence of
considerable quantities of woodland, and even in the Severn Valley
agricultural land seems to have been less extensive. Certainly croplands
are mentioned and there are charter references to flax-lands and wheat-
lands near Worcester, but these are much more rarely encountered on
the township margins, areas where woodland and probably woodland
pasture were still abundant. In addition these parts of the county are
characterised by a higher incidence of enclosures of various sorts,
particularly crofts, worths and grassland paddocks known by the term
gaers-tūn (Figure 11). The charters show that land was enclosed with
living hedgerows and there is some evidence to suggest that the hedges
were primarily connected with the need to keep woodland animals
from pastureland and crops, for the woodlands harboured deer, herds
of semi-wild pigs and even wolves. The prevalence of hedges in an area
of active woodland assarts is particularly noticeable in Central
Worcestershire where the estates of the Church of Worcester extended
into the woodland zone (Figure 12) (Hooke, 1978-9, 3-23). In north-
west Worcestershire there seems to have been less agricultural
development than in the Feldon areas to the south-east and
pastureland and woodland played a more important part in the rural
economy. The emphasis of the Domesday survey of 1086 was almost
entirely upon the agricultural potential of the land and there may be a
danger of underestimating the relevance of other aspects of the
economy in early medieval England. It should be emphasised how well
this evidence complements the evidence of place-names. *Lēah* place-
names, in particular, indicative of woodland country, are infrequent in
the Avon Valley and lands to the south but plentiful in regions to the
north and west (Figure 13). This suggests that the Feldon areas already
presented a more open appearance in the early Anglo-Saxon period.

Systems of Farming Practice

Having ascertained the areas of intensive arable farming in the middle
and later Anglo-Saxon periods it is now possible to enquire whether the

Figure 10: The Vale of Evesham

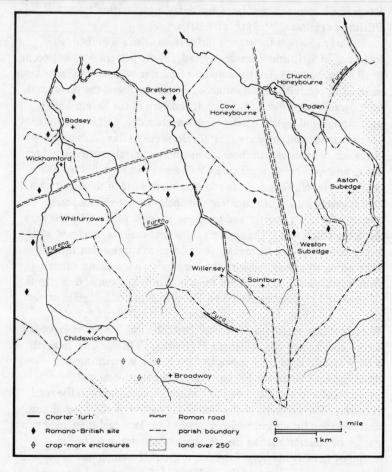

Legend:
— Charter 'furh'
♦ Romano-British site
⏀ crop-mark enclosures
⁓⁓ Roman road
–·– parish boundary
⠂⠂ land over 250'

0 ─── 1 mile
0 ─── 1 km

Figure 11: The Distribution of Features Associated with Enclosures
(i) Croft *Gaers-tūn*, *Edisc* and *Stōd-fald*

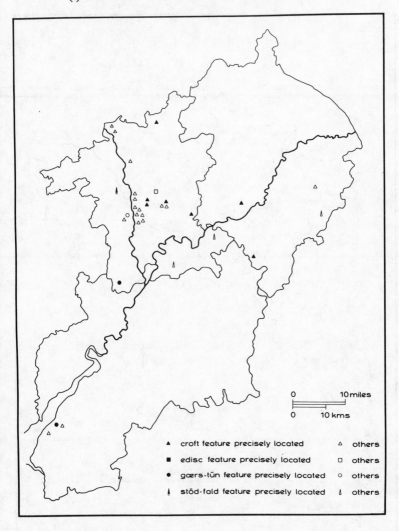

0   10 miles

0   10 kms

▲   croft feature precisely located   △   others

■   edisc feature precisely located   □   others

●   gærs-tūn feature precisely located   ○   others

⌁   stōd-fald feature precisely located   ⌁   others

**Figure 12: The Distribution of Features Associated with Enclosures
(ii) Hedges**

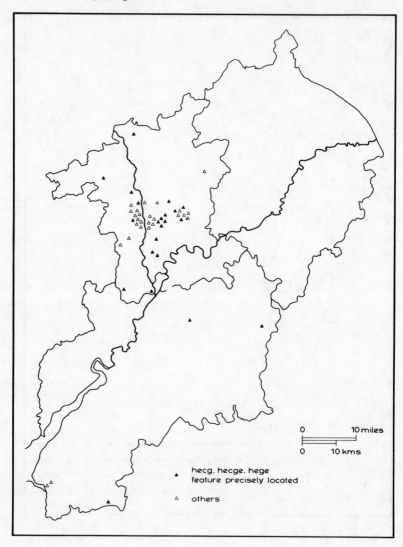

Figure 13: Woodland Areas in the West Midlands Suggested by the Distribution of the *Lēah* Place-name

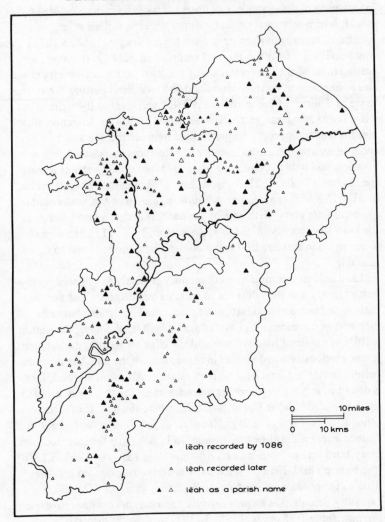

▲	lēah recorded by 1086
△	lēah recorded later
▲ △	lēah as a parish name

charters provide evidence for the systems of farming practice, a question already examined in some detail by Professor Finberg (Finberg, 1972). In fact many of the examples he quotes are taken from the West Midland area. It is in Worcestershire that we may have the earliest reference to communal farming in *tham gemaenan lande* (Sawyer, 1968, S.1272) 'the common lands' of Cofton Hackett recorded in AD 849. However, bearing in mind the interpretation of the term *land,* a community could also hold common rights in woodland and woodland pasture. Thus, the lessee of Thorne, in Inkberrow, in AD 963 was granted the right to cut wood in the *gemaenan grafe* (Sawyer, 1968, S.1305) or 'common grove'. Meadowlands, too, could be partitioned between members of a community and there are a number of references to *dālmāed,* 'meadowland held in common and divided into doles or shares among the holders' (Toller, 1921, 146). A lease of Himbleton, Worcestershire, in AD 975-8 adds that *thonne gebirath se fifta aecer thaere dalmaedue to thaere hide* every 'fifth acre of the partible meadow also belongs to the hide' (Sawyer, 1968, S.1373; Robertson, 1939, 117). Here we seem to see rights in meadowland and woodland held proportional to rights in arable.

Land held in common by a community is clearly in evidence by the tenth century and the reference to partible land suggests that not only pasture and meadowland but arable, too, could be divided between members of a community. Divided arable land is, of course, a common feature of medieval field systems and the charters seem to indicate that it was a well-established system in some areas by the late Anglo-Saxon period. In AD 966 land held with an estate at Alveston, Warwickshire, is described as lying *on thaere gesyndredan hide* (Sawyer, 1968, S.1310) 'in the divided hide' at Upper Stratford, where the lease is of 'every other acre' here and of 'every third acre' of *feld land* or 'open land' at *Fachan leage,* while at nearby Bishopton in AD 1016 the lease includes every third acre of bean-land on *Biscopes dûne* (Sawyer, 1968, S.1388) 'the bishop's hill'. Here we seem to be encountering divided holdings little different from those at Avon, Wiltshire, which in AD 963 are explicitly described as *singulis jugeribus mixtum in communi rure huc illacque dispersis* (Sawyer, 1968, S.719), 'single acres dispersed in a mixture here and there in common land'. Division of strip holdings is further suggested by the phrase *ge inner ge utter,* 'both central and outlying', as in a charter of Moreton in Bredon, Worcestershire, in AD 990 (Sawyer, 1968, S.1363). Here land was divided between two brothers with the proviso that 'the elder shall always have 3 acres and the younger the fourth, both central and outlying, as pertains to the

estate (Robertson, 1939, 133). The phrase *ge inner ge utter* is
encountered in two other Warwickshire charters.

It is noticeable how many of these estates lay in the mid-Avon
valley of Warwickshire or in the 'champion' lands lying to the south,
areas known in medieval times as the Feldon after the amount of open,
cultivated land (Camden, 1586; Dugdale, 1656). This area extended
westwards into the Vale of Evesham and North Gloucestershire. This
does not mean that agricultural land outside the Feldon area remained
untouched by such systems. In Cudley, near Worcester, a tenth-century
episcopal lease refers to *XXX aecra on thaem twaem feldan dal landes
withutan* (Sawyer, 1968, S.1329), '30 acres on the two open share-lands
outside', but in these western regions agricultural land seems only to
have extended as far as the township boundaries in areas near to
Worcester itself.

We have, therefore, a system of divided holdings evident in some
parts of the West Midlands by the tenth century but the origins of the
system are much more difficult to ascertain. Again, terminology may be
relevant. The term *furlang* appears to have developed from *furh lang*
(Toller, 1921), 'the length of a furrow' and to have been from the first
associated with ploughland. The term *aecer* has always been closely
associated with it. Such measured pieces of ploughland are difficult to
envisage outside the context of a carefully measured field system. The
widespread and sophisticated use of these terms in the mid Anglo-
Saxon period seems to suggest that they and their associated field
systems had already by then become firmly established.

Furthermore, I would strongly argue that the significance of the
distribution pattern of agricultural features is one which cannot lightly
be dismissed. The strongest evidence of divided holdings comes from
those areas in which the incidence of agricultural features noted in
charters is highest. It is difficult to disassociate this development from
the pattern of development observed in Roman Britain for these areas
coincided with those in which agriculture is known to have been well
developed in that period (Figure 14). Our knowledge of the period in
the West Midlands is still limited but the distribution of farmsteads
noted from both cropmarks found by aerial photography and surface
pottery scatter strongly suggests that the potential of these same
Feldon areas was fully recognised at an early date. One has to
remember that cropmark recognition is influenced by soil type, while
the ability to find surface pottery scatters is influenced by the
frequency with which land is ploughed and field survey work carried
out. However, extensive fieldwork by the author in west Warwickshire

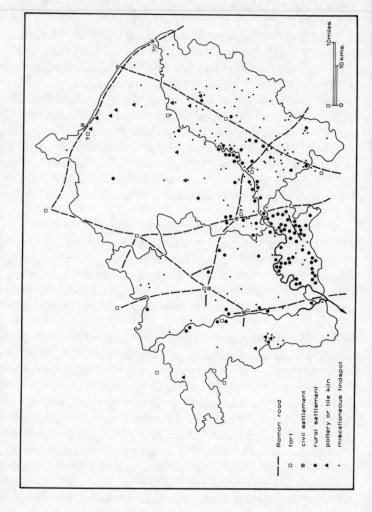

Figure 14: Romano-British Development: the Archaeological Evidence

Roman road

□ fort

▣ civil settlement

● rural settlement

◄ pottery or tile kiln

· miscellaneous findspot

10 miles

10 kms.

indicates that this pattern is not entirely incorrect. There is some evidence to suggest that the same lands continued to be exploited in both the Roman and Anglo-Saxon periods, although the recorded presence of numerous thorn-trees in these areas may indicate some regeneration of scrub over agricultural land in the early Anglo-Saxon period.

Baker and Butlin (Baker and Butlin, 1973, 640) have stressed the influence of population pressure in effecting changes in farming methods. In the present state of our knowledge of Roman Britain it seems that those areas noted for agriculture in Anglo-Saxon times were the very ones in which population pressure was most intense in the earlier period. Recent advances in archaeological knowledge have transformed our notions of population density in Roman Britain and reminded us of the additional economic demands made upon the peasantry under Roman rule (Fowler, 1978, 6). Even the possible increased depredation by grain-devouring beetles may have been underestimated in considering the need for extending the amount of land under grain cultivation (Buckland, 1978, 43-5). It is not unusual for a necessity for increased yields to result in the adaptation and gradual change of farming methods to enable increased demands to be met. Seebohm (Seebohm, 1883, 278) has shown how systems of intermixed holdings appear to have been present in other parts of the Roman Empire by the second century AD. I suggest that the intensity of agriculture noted in the south and east of the West Midlands reflects a pattern of development associated with population pressure already evident in Romano-British times. If the thorn-covered areas noted in the charters reflect some temporary regeneration of scrubland on former cleared land the reversion was probably no more than temporary.

Evidence from other areas shows the same correlation between intensive farming in the Romano-British and Anglo-Saxon periods. In Wiltshire I have noted the frequency with which the *furh* term occurs near field systems in use in Romano-British times. An increasing number of studies has indicated areas in which the boundaries of the medieval furlongs appear to overlie the boundary ditches of earlier field systems. Taylor (Taylor and Fowler, 1978, 159) records a furlong boundary overlying ditches found to contain late third- and fourth-century pottery in Cambridgeshire and there are other similar associations. It need not occasion surprise that early fields might remain visible enough to influence later ones, although this is not, of course, evidence of continuity of land use. Yet in the Feldon areas of the West Midlands there is additional evidence to suggest that in this area at least there is a possibility that a native population continued to farm lands already cleared before the Anglo-Saxon period. There is a mounting body of evidence from pagan Anglo-Saxon burials to suggest cultural overlap and the evidence of place-names only serves to strengthen this supposition. Throughout the West Midland region names containing *walh* and *Cumbra* (Gelling, 1974) seem to imply a British element in the population, and to these must be added the more tenuous evidence of continuity in territorial organisation. With some evidence for communal farming methods including co-aration recognisable throughout the Celtic West the foundations of the open-field system may have been present long before the Anglo-Saxons appeared upon the scene, perhaps within the more limited context of the kinship holding, while the charters suggest that the system was evolved further in the Anglo-Saxon period.

It was also in the areas in which Anglo-Saxon agriculture seems to have been most intensive that the open-field system was to become most deeply entrenched. In the northern and western parts of the area the amount of land under arable in Anglo-Saxon times was noticeably less than in the south and east, with large areas on the western fringes of the Hwiccan kingdom in which development of any kind was minimal. In the medieval period former crofts and wooded areas were to be taken into the extended open fields, as medieval surveys disclose (Hollings, 1934-50) but the intensity of agricultural development seldom reached the proportions noted in the south-east. To the west of the Severn the open fields remained particularly limited in extent. Outlying settlements with their associated lands were able to survive as units and new lands were often available for assarting and for the establishment of additional settlement clusters. Ancient enclosed lands and uncleared land remains to this day (uncleared for arable, that is, but not unmanaged). In the south and east, however, arable land was to be pushed to the utmost margins of the parishes and the open-field system was to become increasingly complex under the pressure of population increase. A growing tendency towards increased settlement nucleation may have taken place and although land was to be continuously reorganised and reallocated, there was little opportunity to change the system for many generations. It is possible that the landscape variations apparent in the medieval period echo patterns already distinct in the Anglo-Saxon period, and that the reasons for these variations should be sought in an even more distant past.

Notes

1. Broadway Inclosure Award, 1771, Shire Hall, Worcester, Af 647.2 (642).
2. D. Hooke, 'Anglo-Saxon Landscapes of the West Midlands: the Charter Evidence' (PhD thesis, Birmingham University, 1980).
3. Throughout this paper the letters 'ae' are used to represent the Anglo-Saxon symbol for which they are a close equivalent.

4 APPROACHES TO THE ADOPTION OF THE MIDLAND SYSTEM

H.S.A. Fox

The field landscape of a typical Midland township, simple and straight-forward though it may at first sight appear today, is the culmination of a number of changes in design which can take us back over at least a millennium and a half of agrarian history, and perhaps even further than that. The last of these changes, which began towards the middle of the nineteenth century, stemmed from the realisation that even the regular allotments planned by enclosure commissioners were not always suited to advances in farming practices. In places the hedges set up in the eighteenth century have been obliterated and with them some other, older impediments to the optimum use of farmland such as ridge and furrow or sinuous boundaries marking ancient divisions within a village's arable territory. Coming before the changes brought about by this 'dis-enclosure' of the field landscape was enclosure itself, a process which has a long history even in the Midlands but which, for many townships there, was completed in the eighteenth and early nineteenth centuries. At this time the enclosure commissioners and the landowners for whom they worked achieved a radical re-orientation of the landscape, planning anew its boundaries though not entirely doing away with traces of earlier patterns. Before this was the arrival of comprehensive open-field systems, a change in organisation which in so many places in the Midlands meant the adoption of a two- or three-field system. How many changes in field structure preceded the arrival of the two- and three-field system is a matter for debate. If (as many historians of the landscape are now coming to believe) the two- and three-field system was a culmination of many centuries of cultivation, then several earlier stages must be envisaged. One of these may have been characterised by a field system in which much of a township's arable was already divided into strips yet which lacked organisation on a village basis or any division of the land into two or three great sectors or fields. Earlier, in the first centuries after the first English settlements, there seem to have existed systems which were less extensive than these and which more closely resembled farming in severalty, some legacies of which survived subsequent changes and may be identified on pre-enclosure maps and even in the landscape today. Some would argue that, before this, both logic and the archaeological evidence demand a substantial Celtic contribution to the history of the field landscapes of Midland England.

It is becoming increasingly clear that the arrival of the two- or three-field system came somewhere near the centre of a progression of stages in the evolution of patterns of cultivation in the Midlands and that each stage has left behind some traces, however faint, in the landscape of today. The arrival of the system had a deep influence on the subsequent agrarian history of the places where it took root: on their agrarian landscapes — for the nature of the system tended to prolong its life and was inclined to mean that its abolition should take the distinctive form of enclosure by statute — and also, of course, on their subsequent social and technological development. Equally, an understanding of the arrival of the system should provide the firm basis which is needed for an exploration of earlier field and farmstead groupings in Midland townships. This paper is a survey of the previously published work (and a small sample of the types of documentary evidence upon which it has been based) bearing on the dating of and processes behind the adoption of the two- and three-field system. It is ground which has been covered before on several occasions but another general survey may, nevertheless, assist new detailed studies — based upon both documentary and topographical evidence — which will eventually enable questions about these subjects to be answered with greater confidence.[1]

The Essential Features of the Midland System

The two- or three-field system, or Midland system, was what agricultural historians still call an 'open-field system' in the sense that its smallest units were unfenced, 'open' strips. These strips were normally grouped into interlocking bundles, each called a *cultura* in the Latin terminology of medieval documents and, in the vernacular, described by a variety of terms of which 'furlong' was probably the most frequently employed and is that used by most historians today. The furlongs were grouped into two or three great sectors almost invariably called 'fields' or *campi* in the documents. The fields of a township, whether two or three, were of approximately equal sizes and the strips of a normal holding were distributed approximately equally between them. Each year one of the fields was set aside as common fallow grazing for the animals of all cultivators; regulation of grazing in the fallow field, as well as regulation of other aspects of husbandry, was achieved communally by those who had a stake in the system. This was the agrarian regime which

H.L. Gray in 1915 called the Midland system because its home territory in England was a broad belt or zone of land running diagonally through the centre of the country from mid-Somerset in the South West to lowland Northumberland in the North East, although naturally there were many places within this area which used other types of arrangement. This, too, was a system akin to the 'common fields' discussed by Joan Thirsk in 1964, although her initial definition did not insist upon equality in the sizes of the fields.[2]

It should at once be clear from this simple description of the Midland system that only two features were unique to it, marking it out from all other types of agrarian regime. Subdivision of the arable land of a settlement into unfenced strips lying in furlongs was a characteristic of many types of system; most — but not all — variants of 'open-field' systems involved the common grazing of furlongs or other patches of arable, although all patches were not necessarily common to all cultivators; some degree of communal regulation or overseeing was a feature of almost all field systems.[3] But unique to the Midland system was a grouping of the furlongs into two or three compact fields of roughly equal size among which the arable strips of holdings were more or less equally apportioned. Unique also was the allocation each year of a compact block of land, one of the fields, as common grazing for the animals of all cultivators.

It should also be clear that these two unique features of the system were closely related to one another: the second, functional feature found its expression, in terms of field layout and of the landscape, in the first. It is difficult to envisage how one-half or one-third of a community's arable could have been efficiently grazed in common by the flocks and herds of all cultivators without a division of the land into two or three great compact fields. If such comprehensive grazing were introduced into an arrangement where the fallow patches were scattered about in small blocks among cultivated land, any one of the blocks would not be able to support all of the community's animals for the whole of the fallow season, necessitating frequent and inconvenient movement of livestock or an equally inconvenient allocation of animals among the blocks. Conversely, it is difficult to envisage what motive lay behind a division of the arable of a township (and of each holding within the township) between two or three fields if it were not a desire to set aside each year a new compact half or third of the land for fallow grazing. It cannot be argued that a two-course rotation necessitated a two-field system and a three-course rotation a three-field system. There were numerous townships in medieval England which practised deviations, within a framework of three fields, from classical three-course rotations in which each cropping unit was subjected to a regularly repeated three-year cycle of land use. So too were there numerous townships with field systems in which a three-course rotation was practised without finding expression on the ground in three great fields. In these cases there is much truth in Sir John Clapham's dictum: 'Crop rotation is independent of the lay-out of the fields' (Clapham, 1949, 54).

It is worth emphasising this point with some examples, for they will serve to underline the fact that the essential features of the two- and three-field system were determined by the requirements of fallow grazing rather than by the requirement of particular crop rotations. In Cheshire, where there is not abundant evidence for the Midland system, the demesne lands of each of the two manors of Wybunbury and Tarvin were divided into three 'seasons' according to an extent drawn up in the late thirteenth century; but these three seasons did not take the form of three great compact fields. Typical was one of the seasons at Wybunbury which comprised three small crofts, a larger field called 'Clauerthyn' and part of an assart called 'Vicariesruding', other portions of which were assigned to the other two seasons. It was not a compact block but an 'artificial' collection of several pieces of land so that they might be assigned to the same course in the cropping sequence (Sylvester, 1959; 1969, 241-2). In an adjacent county, the survey made in 1334 of the Honour of Denbigh shows, again, how a three-course rotation did not need three great compact fields for its operation (Vinogradoff and Morgan, 1914, 1, 2, 4, 230). On two of the manors the demesne had been, in the words of the survey, 'converted into three seasons', almost certainly as part of a reorganisation of the estate after it had been taken out of the hands of the Welsh, for we are told that one of the manors had been 'made' by its first English owner, Henry de Lacy. But 'conversion' had not involved any change in field plans; rather, it seems that pre-existing blocks of land which had formed the field system before the English annexation had for cropping purposes been grouped — not physically but in the minds of officials — into three seasons. On a third manor the grouping had not yet been made in 1334, but the survey states that it could well be (*tamen potest bene tripartiri*): this statement does not imply potential for the recasting of the structure of the fields into three[4] but simply records the potential of the manor for use of a three-course rotation. Many other examples could be cited, from manors outside the zone of Midland field systems, of three-course rotations which operated without the existence of three fields.[5] In such cases the unit of rotation was often described as a 'season'[6] and the framework of fields was often what will be called in this paper, adapting Gray's terminology (Gray, 1915, 83), an irregular multi-field framework. Townships with systems of this kind had assigned each of their many fields and furlongs to a particular season in order to facilitate cropping, but had not experienced the need to introduce comprehensive fallowing arrangements. The former action had required no alteration of the layout of the land; only the latter, if adopted, would of necessity bring with it a reorganisation designed to create large, compact fields suitable for common grazing on a village-wide basis.[7]

To have isolated two essential and closely-related characteristics of the Midland system has important implications for our search after its origins. In the first place it suggests that the origin of the system was closely connected with a need to find grazing land for the livestock which were so important in ensuring that the land was ploughed and the ploughed land fertilised. In the second place, it will make a search for origins marginally easier, for it so happens that the earliest documents surviving in some number *and* containing some details of fields are those medieval charters from the middle of the twelfth century onwards which give information about the manner in which the parcels of a holding were distributed but which contain very little other evidence of agrarian arrangements. Once it is realised that an equal division of the strips of a holding between two or three fields was a diagnostic feature of the Midland system, implying biennial or triennial fallowing of whole fields, then a search among twelfth-century documents for indications of the presence or absence of the system becomes more simple.

Conjectures on Origins

H.L. Gray was the first to scrutinise carefully a large sample of evidence relating to English field systems. He was without detailed evidence on the field systems of the Saxon period and therefore assumed that 'distinctions which obtained in the thirteenth century are assignable to the period that saw the accomplishment of Saxon settlement'. He also assumed — following Meitzen — that in their homeland the invaders already practised a system akin to the Midland system. His evidence appeared to show that the area dominated by the Midland system broadly coincided with the area where Saxon settlement had been strongest and he was thus able to conclude that its introduction there was further proof of 'the thorough Germanization of central England' (Gray, 1915, 411). His assumptions (reasonable enough in their day) that the two- or three-field system was an early and an Anglo-Saxon system were also his conclusions.

Gray's conclusion that the Midland system was a familiar feature in the early Saxon period would by no means have seemed exceptional to contemporary scholars. Nasse, with his idea of a 'primitive' Germanic village community, Seebohm, intent upon tracing English rural institutions to Roman or pre-Roman roots: both had taught their readers to expect the system in the fifth century, if not earlier (Nasse, 1871, 14-26; Seebohm, 1905 edn, 410-11). They were both, however, like Gray, content to avoid giving even a poorly focused picture explaining how and why the system might have been adopted by early farming communities. It was not until 1938 that such a view was first provided, by C.S. and C.S. Orwin in *The Open Fields*.

In contrast to the Harvard scholar, the Orwins were intent on applying the historical imagination, based upon their deep acquaintance with the land, to the question of how pioneer settlers in the past might have grappled with problems of survival. Their model is as conjectural as Gray's, but it is based less upon assumptions about continuity between the fifth century and the thirteenth and about the migration of material culture, more upon suppositions about the perceptions and responses of farmers under pioneer conditions. It portrays groups of settlers who were driven by the need for survival to co-operate in bringing their newly acquired woodland or waste into cultivation and, as a result, divided it among themselves into narrow strips and managed it according to common rules. Believing that a biennial rotation, with a fallow every second year, was common-sense practice for any settled group of farmers, the Orwins supposed that these strips would have been cast from the first into two fields; at a later stage the two fields might have been made into three. A product of communal clearing, the arable land would naturally have been turned over to common grazing when it lay fallow (Orwin and Orwin, 1967 edn, 39-41, 53-5). The Orwins thus neatly side-stepped a cultural explanation for the Midland system.[8]

Persuasive, eminently practical and down-to-earth, the picture given by the Orwins became an orthodoxy. With small changes in detail, it was accepted, for example, by F.M. Stenton in 1943, by W.G. Hoskins and D. Whitelock in 1955 and, more cautiously, by H.R. Loyn in 1962 (Stenton, 1943, 277; Hoskins, 1955, 38; Whitelock, 1955, 70; Loyn, 1962, 160-3). It was not subjected to detailed scrutiny until 1964 when Joan Thirsk looked anew at some of its basic assumptions.

Thirsk's model may be described as evolutionary rather than cultural or functional, for one of its basic premises is that it is not necessary to assume that all features of a field system in its maturity have always existed alongside one another. Those parts of her model which deal with the genesis of strips and furlongs need not be considered here, for what we are concerned with is the origin of a whole system of management, 'the common fields', which she describes at the beginning of her paper as comprising 'arable . . . thrown open for common pasturing by the stock of all the commoners after harvest' and 'the ordering of . . . activities . . . by an assembly of cultivators'. She states further on that 'the grazing of all the fields of a village in common could not take place until they were all incorporated into a scheme of cropping which assured that all the strips in one sector lay fallow at the same time', making it clear that the 'common fields' of her discussion was similar in essence to the Midland system of Gray (Thirsk, 1964, 3, 18-19).

According to the new model the system originated through the refashioning of earlier, irregular multi-field arrangements in which a large proportion of a community's arable had already become divided into strips but which lacked the other features of the Midland system. In their form, these earlier field systems exhibited a 'puzzling appearance of disordered cultivation', each township having 'numerous fields, not apparently arranged in any orderly groups'. As far as pasturing was concerned, agreements were made 'between neighbours possessing intermingled or adjoining land' (Thirsk, 1964, 16, 19). But the great, compact fallow field was absent in both concept and practice. The stimulus which encouraged remodelling of systems of this kind was a growing need, as communities grew and waste diminished, to use fallow as efficiently as possible, a need which brought them to a point at which it was necessary to introduce a system 'in which all tenants shared common rights in all fields'. In the interests of the viability of individual holdings, a 're-distribution of land was necessary in order to facilitate the introduction of new common-field regulations' (Thirsk, 1964, 15, 18, 20). This remodelling of field systems might be achieved at a stroke or through exchanges made over a number of years, processes which took place 'at different times in different parts of the kingdom' (Thirsk, 1964, 22, 24). Thirsk concludes that, in general, 'rights of grazing over arable land were still being shared by neighbours in the twelfth century, but before the middle of the thirteenth century there were villages in which all tenants shared in common rights in all fields' and that 'we can point to the twelfth and first half of the thirteenth centuries as possibly the crucial ones in the development of the first common-field systems' (Thirsk, 1964, 18, 23).

This thesis has been referred to here as a 'model' because it is impressive as a logical argument, supported by comparative evidence, about how a system like the two- or three-field system might have emerged at some time in Midland England. It has coherence and elegance; moreover, in linking the emergence of mature field systems to a need for efficient use of fallow it explains, more satisfactorily than alternative models do, the essential features of the Midland system. But as an account, supported by evidence *in situ* as it were, of what changes in agrarian structures were taking place in the Midlands in the twelfth and thirteenth centuries it can be demonstrated to be imperfect.[9]

Novel interpretations intended 'to open . . . fresh discussion' (Thirsk, 1966, 142) can pass too easily into the literature as more or less accepted orthodoxy. Thus the latest general works on English field systems and on rural society in medieval England, excellent though they are, appear to accept a post-Conquest date for the origin of mature field systems in the Midlands without questioning or adding to the evidence upon which Joan Thirsk's presentation was based. To C.C. Taylor, 'it is doubtful whether the system had actually evolved sufficiently to reach this [classic] state much before the twelfth century'. To E. Miller and J. Hatcher 'there can be little doubt that the time of definition in many places did lie in the twelfth and thirteenth centuries', although they admit to the possibility of an earlier origin in some townships. H.E. Hallam has described 'the creation of fully mature common fields' as a major change 'which took place in the open-field system after 1086'; R.C. Hoffmann has claimed that 'emphasis must be placed . . . on the late twelfth and thirteenth centuries as a time of unusually pronounced increase in the extent and severity of communal restrictions on individual management of arable resources'. More cautious was the assessment of A.R.H. Baker and R.A. Butlin. Noting how many examples may be found of changes in field systems in the centuries after 1300, including examples of wholesale remodelling of field plans, they argue that 'it becomes on *a priori* grounds alone somewhat dubious to postulate their inflexibility before 1300 and since the time of the original settlement' (Taylor, 1975, 72; Miller and Hatcher, 1978, 96; Hallam, 1972, 219; Hoffmann, 1975, 49; Baker and Butlin, 1973, 650).

Alongside this widespread acceptance of the revisions suggested by Joan Thirsk must be set the severe criticism of J.Z. Titow. Close acquaintance with medieval documentary evidence bearing upon the problem, as well as other lines of argument, lead him to ask if it is not really 'much simpler to envisage an original settlement which created the nucleus of the regular elements – the standard holdings, the strips, the even distribution among the rotational fields – and which was expanded . . . by succeeding generations of villagers'.[10] This question appears to take the argument back to its beginning, to the ideas of Gray and the Orwins. It is time, therefore, to look again at some of the evidence for the chronology of field systems in the Midland belt.

The Chronology of the Midland System

Three principles must dominate any attempt to establish from documentary sources a tentative chronology for the appearance of the Midland system. First, the historian must be quite clear that he knows what he is looking for: he must be aware of the essential and diagnostic features of the system and of how these appear in the documents. Second, he must understand thoroughly the chronology of the documentation, so as not to run the risk of writing the history of charters and courts rather than of fields and farms. Third, he must try to understand when he can expect to find references to the system as a matter of course, when the references which perchance he finds are exceptional in the documents but do not refer to exceptional phenomena, and when he can expect to find no references at all.

The Period c. 1250 - c. 1350

It is essential, in light of recent arguments, to establish at what point in the documentary record the Midland system may first be observed in all of its details. This is important because one repeated criticism of Joan Thirsk's suggestion of a post-Conquest origin for 'the common field system' is that she insists upon identifying features which, critics claim, were inessentials of the system, not appearing in the documents until a relatively late date. Titow complains that her definition emphasises 'accidentals' and portrays the system 'as it was in the late sixteenth or seventeenth centuries, that is, at the earliest time when documents permit us to see it in its entirety' (Titow, 1965, 89). His view is repeated by M.M. Postan: 'The common field system could be defined very precisely to apply only to cases where it happened to be uniform, wholly symmetrical and above all comprehensive . . . Very few of the manorial and village fields in the early middle ages would satisfy a definition as rigid and restricted as this. Historians adhering to the definition would therefore be quite justified in concluding that field systems rigorously and symmetrically organized . . . were a late medieval or even early modern product of a deliberate administrative reform' (Postan, 1972, 56). Two questions must therefore be asked. The answer to the first – were symmetry and regularity minor or essential features of the system? – has already been given in an earlier section of this paper: far from being accidentals, a symmetrical division of the arable and of individual holdings and the regular allocation for grazing of a fallow field were essentials which lay at the very heart of the Midland system. Second, is it really in the sixteenth and seventeenth centuries, or in some earlier period, that these essentials may be observed in detail?

Descriptions of demesne arable in manorial extents from the period between 1250 and 1350 provide many clear examples of one essential and highly regular feature of the Midland system, namely an equal division of the land between two or three great fields. No attempt can be made here to marshal all the evidence of this kind, but Table 4.1 (page 80) gives details of three estates, each including some manors within the belt of Midland field systems. For the first estate, the Glastonbury Abbey lands in Somerset and North Wiltshire, I have used the extents drawn up in the early fourteenth century;[11] for the second estate, that of the Bishopric of Ely, I have used the detailed survey made in 1251; for the third, the estate of the Bishopric of Coventry and Lichfield, extents were drawn up towards the close of the thirteenth century.[12] The table shows demesnes which were for the most part regularly apportioned between two or three fields, the acreages deviating from a 'perfect' bipartite or triparte division by only a few per cent in most cases. On some manors the regularity of the disposition of demesne between fields was remarkable: so careful had been the planning of the fields that extensive demesne holdings of several hundred acres had been divided into two or three parts whose sizes differed by no more than a few acres.

Much evidence for a similar degree of symmetry could be cited, from some manors on the Fortibus Yorkshire estate, for example, or from the lordships of the bishops of Worcester and the bishops of Lincoln.[13] It could be objected that extents do not normally specify that demesnes were intermixed among the strips of peasant holdings and that some descriptions of demesne lands divided between two or three *campi* could have applied to two or three great compact demesne closes lying separate from the land of tenants. This objection could be examined in every case only by detailed scrutiny of other types of document less convenient to handle and to summarise than the extents; suffice it to say that throughout two- and three-field country it was by no means unusual for demesne lands to be intermixed with the land of tenants[14] and that on one of the estates covered in the table this can easily be shown to have been so.[15] It could also be objected that abnormal estates have been selected, with demesnes whose neatness and symmetry reflect the strength of wealthy central administrations presided over by men with the administrative calibre of a Bishop Grosseteste or a Henry de Blois; and that it has not been shown that the land of peasant holdings shared the same symmetrical division between fields. Both objections are ruled out by other sources, particularly the evidence of countless charters from the period between 1250 and 1350 which describe, for manors on estates of all types and sizes, peasant holdings whose arable land was distributed equally between two or three fields.[16]

In sources from between 1250 and 1350 a second regular feature of the Midland system may be observed, namely the allocation each year of one of the fields as fallow grazing which was the dominating principle of the system. It has now become customary, almost fashionable one might say, to stress how flexible the Midland system was in terms of cropping practices. R.H. Hilton, who was one of the first to make this suggestion, found that 'in some villages' of medieval Leicestershire 'field divisions were ignored, and the furlong regarded as the basic unit for purposes of cropping' (Hilton, 1954, 160). But his important conclusion about the flexibility of medieval cropping has been taken up in general statements which seem almost to do away with any connection between the form and the functioning of the Midland system: 'the unit of rotation was often not the field but the furlong'; 'the existence of two, three or more common fields had no significance at all in agricultural terms' (Baker, 1965, a, 88; Taylor, 1975, 72). This is going too far. *Of course* the Midland system was flexible in terms of cropping and allowed for deviations from rotations in which half the sown land in any one year carried spring crops and half carried winter crops. Changes in consumption habits, in market conditions and in harvest qualities all dictated that cropping on this basis was exceptional rather than normal by the thirteenth century, and may have always been so. The furlong rather than the field was thus the unit of cropping, but from all accounts it appears that the field retained a central place in the whole system of rotation: whatever changes in cropping were rung on the furlongs of the sown field or fields, the fallow field remained inviolate. It is not easy to support this contention as strongly as critics might wish, for students of field systems have tended perversely to ignore the only medieval sources which give full information on rotations, namely cropping plans which occur in the most detailed of manorial *compoti*. Most analyses of cropping plans which show an absence of regular fallowing of whole fields come, as we might expect, from townships which lay away from areas characterised by Midland field systems.[17] But the evidence, analysed so far, from townships with Midland field systems, points conclusively to the integrity of the fallow field. At Cuxham (Oxfordshire) one or another of the three fields remained unsown in each of the 50 years between 1297 and 1358 for which the relevant details have survived; when there are sources for consecutive years they show each field returning to fallow every third season. At Bunshill (Herefordshire) between 1326 and 1344 one or another of the three fields was always absent from the accounts, indicating that it lay fallow. For Glastonbury Abbey's two-field township of Walton (Somerset) *compoti* have been examined for 17 years between 1331 and 1477, revealing that a variety of crops was grown on the sown field in any one year but that the other field was invariably left fallow with the occasional exception of a few acres used for beans or barley. The Abbey practised an identical regime at Buckland Newton and Plush (Dorset) in the early fourteenth century. For each of three Oxfordshire townships, D. Roden has stated that 'manorial accounts . . . show that the whole of one field lay fallow each year'. With small variations the same conclusion may be reached for Landbeach (Cambridgeshire) in the middle of the fourteenth century, while a fallow field appears as a constant feature of the working of the three-field system at Oakington, in the same county, in most of those fourteenth-century seasons for which evidence is available. Hilton's analysis of cropping plans in the *compoti* for Kirby Bellars (Leicestershire) for three years in the fifteenth century shows that in each of these years either the Westfield or the Middlefield or the Eastfield was strictly reserved for fallow.[18] This is a small sample, and until it is supplemented by further work on *compoti*, other more circumstantial types of evidence must be used to reinforce the impression that the Midland system was less flexible in terms of fallowing arrangements than it was in terms of cropping.[19]

A third feature of the Midland field system which we may observe with reasonable clarity in documents from between 1250 and 1350 is the regulation of aspects of husbandry by meetings of villagers. These meetings were the assemblies of cultivators whose activities, although normally leaving no documents of their own, were on some manors set down from time to time in the records of the lord's court. The regulations which they enacted were described in the vernacular as by-laws, 'the laws of the village'. Early by-laws concerning the management of fields are rare, for many of them undoubtedly went unrecorded in writing and because court rolls for the years before 1300 are not common. But enough examples may be cited from W.O. Ault's excellent collection of references to such laws and their infringement for us to be able to observe, at townships in many parts of the belt of Midland field systems, the detailed and minute regulation of village fields in the years around 1300.[20] Harvesting was speeded by enactments which prevented diversion of the casual labour force of the village to gleaning before all the grain was carried, an important provision if the harvest was not to be spoiled; other by-laws regulated the transition between the use of a field for grain and its function as pasture, stipulating, for example, that animals should not be tethered on a furlong until adjacent furlongs had been cleared of crops; early by-laws from around the year 1300 record, too, the progression of animals into the fallow field, mares tethered with their foals, then cattle, then the village flock; when the fallow field was again emptied of stock in preparation for sowing, there were by-laws to ensure that it was securely fenced. Early agrarian by-laws touch upon many other aspects of the farming year, but Ault has acutely observed that 'by-laws of harvest were the ones most commonly recorded by thirteenth- and fourteenth-century scribes' (Ault, 1965, 40). The village officials in charge of their enforcement were thus sometimes termed *custodes autumpni:* this was the anxious time of transition between a field's use for crops and its essential deployment as a pasture for stock, the juncture which was at the heart of the Midland system.

In sum, extents, *compoti* and court rolls leave no doubt that it is in the latter half of the thirteenth century and first half of the fourteenth — and not in the sixteenth or seventeenth centuries — that the Midland system may be observed in all of its details. The picture drawn here may to some appear to be reactionary, closer to a textbook diagram than to reality, more a caricature than a faithful portrait. The sources do not bear out such a view. Local variations in topography of course gave the system an expression on the ground which varied remarkably from place to place and on some manors they gave an apparent anarchy to the arrangement of the fields. Nor was it an unchanging system, for it could accommodate changes in crop combinations, new assarts and exchanges of land without losing its essential elements.[21] But there can be no doubt that the system was widespread and in perfect working order around the year 1300; moreover, the elements of symmetry and

regularity which it displayed suggest perhaps that it was then a more
recent innovation than some would allow. When we come to ask how
recent it was we must turn our attention to different types of source material.

The Period c. 1150-c. 1250

Twelfth-century surveys are far fewer than those from the middle of
the thirteenth century onwards and in their descriptions of a demesne
they normally confine themselves to laconic statements of its value or
the number of teams needed for its cultivation.[22] The earliest *compoti*
to contain cropping plans appear to date from the second half of the
thirteenth century; no *compoti* are known from the twelfth century.
Likewise, the evidence of by-laws fades before the 1260s and 1270s:
very few court rolls earlier than this have survived. The general history
of estate management and administration combines with the histories
of survival of records from particular estates to make the middle of the
thirteenth century a turning point in our detailed knowledge of field
systems. For an earlier period we must rely, for written evidence,
almost exclusively upon charters.

Charters vary greatly in the amount of information which they give
away. Three broad groups may be recognised. First, there are charters
which contain virtually no useful information about field systems,
referring simply to a specified number of acres, to a virgate (or one of
its equivalents, or multiples of divisions of these units) or to 'all' of the
land of a grantor in a particular place. Such are two charters dating
from the first decades of the thirteenth century by which land in the
manor of Rodmarton (Gloucestershire) was granted first by Philip son
of Reginald to Ralph of Gloucester and then by Ralph in pure and
perpetual alms to Cirencester Abbey. The land was simply described as
a virgate, and its former owners recited. But an earlier, twelfth-century
charter granting the same virgate to Philip son of Reginald — possession
of which by subsequent owners of the land no doubt released them
from the need to specify its layout in later documents — is far more
detailed. It, too, names the former owners of the virgate but, in
addition, gives locational details: *he vero sunt acre predicte virgate
terre*, a statement which is followed by a long list of the acres making
up the virgate, half of them (22½ acres) in the north field and half
(23½ acres) in the south field. If only the later charters relating to this
virgate had survived we would have had no means of knowing that it was
a perfect example of a holding forming part of a two-field system.[23] A
second group of charters provides more specific details: these are
charters naming the furlongs within which the land lay but giving no

indication as to whether or not they were divided between two or three great fields. Their interpretation sometimes presents difficulties which may be illustrated by the example of a grant made to Osney Abbey in 1303 of 12½ acres in the manor of Hampton Gay (Oxfordshire). The grant names a number of furlongs — 'Piteneyesforlong', 'Goldebroke', 'Banlonde' and so on — but makes no mention of any larger divisions or fields. Yet it is clear from other documents relating to Hampton Gay copied into the cartulary of Osney — including some twelfth-century charters and an unusual and early list of lands which owed tithe to the church of St George — that most of the furlongs of the 1303 grant lay within the two fields of the village (Salter, 1929-36, vol. 6, 74, 39-76). The grantor in 1303 made no mention of this fact, perhaps because he thought it more precise to locate his land by furlong than by field. Villages which appear from charters to have possessed numerous ungrouped units in an irregular multi-field pattern may therefore, in some cases, have been two- or three-field villages.[24] The reverse, however, does not apply: charters in a third class, which specifically state that the land lay in two or three fields (and indicate an approximately equal division between them) can only be taken at their face value. Some charters of this type both name the furlongs and also indicate their grouping into fields; others refer to the fields alone, a practice whose precise purpose is uncertain, for it can hardly have helped the grantee to locate his newly acquired strips (yet it was common enough practice: might the idea behind this terminology have been to reinforce the grantee's claim to a stake in a *system* of husbandry involving the two or three great fields?)[25] References to the fields themselves come in a variety of forms: so many acres *in uno campo* and so many *in alio* is a common formula; or the fields may be given their names; or reference may simply be made to a holding with land *in utroque campo*. Whatever the formula, there can be no doubt that it had a real meaning and expression on the ground.

Such references, which begin to be included in charters dating from about the middle of the twelfth century, are the oldest references which we have to holdings equally divided between two or three fields.[26] Because a division of this kind implies (as argued above) management on a two- or three-field basis, such references furnish our earliest explicit evidence of the Midland system. And because references to field arrangements in charters survive in relatively large numbers they provide a far surer basis for an assessment of the presence of the system between 1150 and 1250 than may be obtained from much rarer and exceptional references, not the normal concerns of land grants, to

common rights over the arable lands of a vill (Stenton, 1920,
xliii). Nor may it reasonably be objected that the evidence of field
arrangements in charters is 'almost irrelevant to our discussion', or
relates to land of 'unusual character' (Titow, 1965, 98-9). Most
collections of early charters contain grants of complete and perfectly
normal holdings both free and unfree, as well as grants of portions of
demesnes which grantors may have selected for their typicality of a
whole field system. Sir Frank Stenton put the matter most clearly when
he described early grants 'which illustrate the arrangement of open
fields' as 'precious' and of 'especial importance' for the study of
agrarian history (Stenton, 1922).

A brief survey has been made of some of the more easily available
charter evidence, and of interpretations of it, from two areas within the
belt of Midland field systems. For the first area, Gloucestershire,
students of field systems are lucky in having in print three monastic
cartularies containing many early grants to religious use: the great
'Landboc' of the Benedictine foundation at Winchcombe, the
cartulary of St Peter's Abbey at Gloucester and two registers of title
deeds and other instruments compiled at Cirencester Abbey.[27] In
addition to the Gloucestershire properties mentioned in these three
cartularies, parts of the county have been covered in the admirable
parish-by-parish surveys of recent volumes of the *Victoria County
History of Gloucestershire*. These incorporate excellent thumb-nail
sketches of field systems, including references from twelfth- and early
thirteenth-century charters when they are available. They show,
incidentally, how common it is for a parish to have no telling
references to its fields until the seventeenth or even the eighteenth
centuries. These two main sources for Gloucestershire have been
supplemented by a small number of other references. For the second
area, those parts of Cambridgeshire (including the Isle of Ely) and
Lincolnshire which fell within the two- and three-field zone, two recent
discussions of field systems have made use of early charters. M.R.
Postgate examined evidence, from charters and other sources, on the
field systems of every Cambridgeshire parish, while much of the
published early charter evidence from Lincolnshire has been summarised
by H.E. Hallam.[28] These two works have been used extensively here,
and examples from a few other sources added. For Gloucestershire the
findings are as follows: evidence of the Midland system before 1250 has
been found for approximately 30 townships, an impressive figure
considering the nature of the sources and the fact that parts of the
county – the Forest of Dean, Severnside and the Cotswold edge – were

Table 4.1: Demesne Holdings in Two- and Three-field Townships on Three Medieval Estates

Estate	Township	County	Fields (acres)		
			1	2	3
Glastonbury	Shapwick	Som.	256	245	–
Glastonbury	Greinton	Som.	65	68	–
Glastonbury	Ashcott	Som.	163	152	–
Glastonbury	Street	Som.	241	249	–
Glastonbury	Meare	Som.	76	56	–
Glastonbury	Pennard	Som.	143	134	89
Glastonbury	Wrington	Som.	165	154	–
Glastonbury	Milton	Som.	92	102	81
Glastonbury	Walton	Som.	159*	164*	–
Glastonbury	Nettleton	Wilts.	134	136	–
Glastonbury	Grittleton	Wilts.	196	165	–
Glastonbury	Kington	Wilts.	118	127	–
Glastonbury	Christian Malford	Wilts.	81	156	–
Glastonbury	Winterbourne Monkton	Wilts.	261	289	–
Ely	Littleport	Cambs.	100	100	80
Ely	Downham	Cambs.	160	150	134
Ely	Wilburton	Cambs.	96	72	108
Ely	Linden	Cambs.	240	280	249
Ely	Willingham	Cambs.	82	91	106
Ely	Triplow	Cambs.	108	153	111
Coventry & Lichfield	Bishops Itchington	Warwcs.	206	216	–
	Chadshunt	Warwcs.	242	254	–
Coventry & Lichfield	Sawley	Derbys.	105	124	93
	Eccleshall	Staffs.	259	311	241

*Leased acres excluded.

never dominated by two- or three-field arrangements. For Cambridgeshire and Lincolnshire the evidence which has been examined yields approximately 20 and 65 townships known to have practised the Midland system before 1250. The high figure for Lincolnshire is in part a product of abundant early charter evidence, in part a reflection of the fact that, the Fenlands excluded, Midland systems were remarkably widespread within the county, alike in the Wolds as in the lower claylands. The evidence upon which these figures are based is set out in an appendix at the end of this paper.

In view of these findings it is difficult to support the contention that the twelfth century and the first half of the thirteenth marked a crucial stage in the development of the Midland system. It might be argued that the number of places which have been counted here – those for which early sources are available and for which these sources indicate a two- or three-field system — is relatively small. But this would be to have too great an expectation of twelfth- and early thirteenth-century charters. Indeed, historians who have worked with such sources might conclude the opposite: that although the numbers can never be expressed in terms of a mathematical probability, they nevertheless provide impressive evidence for a system which was already commonplace and well established by the end of the twelfth century. One must agree with Titow's assertion that farming 'on the basis of two- or three-field systems was known on tenant land . . . in the twelfth century' (Titow, 1965, 98). Had the system been a novelty then we would expect to find far fewer references to it in the reign of Henry II. We would also, perhaps, expect to find far more references than we do to the processes of its arrival.

This last line of argument is not an easy one to support, but it is worth taking as far as possible. Most of the cases which are often cited as examples of the wholesale remodelling of fields have little direct bearing on the adoption of the Midland system. Some, such as F.M. Stenton's suspected case of 'rearrangement of . . . fields' at Great Sturton (Lincolnshire) in *c.* 1150 or R. Lennard's from Bedfordshire in 1304, turn out to be examples of far more common procedures: the former simply involved a consolidation of a demesne holding and the latter seems to have been an 'inhok' in one of the common fields.[29] Others relate to a change from a two- to a three-field system, as occurred, for example, at Puddletown (Dorset) where in 1291-2 the bailiffs of the manor 'bounded the fields in three parts' (of 168, 177 and 175 acres)[30] or at Mursley and Dunton (Buckinghamshire) where in 1345 it was agreed that 'in order to take greater profit from the lands . . . the tenants . . . may in future sow two parts of all their lands each year'.[31] Yet others refer to adjustments within three-field systems, such as the case made famous by Joan Wake where in the later Middle Ages the community of the vill of Harlestone (Northamptonshire) agreed upon a redefinition of the fields and an enlargement of the access roads (Wake, 1922). There are also examples of the replacement of two or three fields by four, a common enough development in the seventeenth century and one which almost certainly has some antecedents in the later Middle Ages.[32] Finally, and less easy to assess, is the case of Marton (Yorkshire) where the two lords of the vill and its freemen elected nine representatives to divide the land 'so that one part every year be fallow'. The date, 1344 x 1403, is relatively late and no evidence has yet been found to show that this was not an example of a change from a two- to a three-field system.[33]

All of these cases are of great interest in so far as they show, very clearly, that communities could come together in order to adjust their field systems and, less precisely, some of the means which were used in such complex operations. But they contribute nothing to the notion that the Midland system was an innovation of the twelfth and thirteenth centuries. It has been suggested that many communities might have arrived at regular field systems not through a single wholesale redistribution of the land but by a series of exchanges, spread over a number of years, which would have left fewer traces in the documentary record (Thirsk, 1964, 22; cf. Thirsk, 1973, 259-60). This may have been true of some places but the equality of distribution of holdings between fields in so many Midland townships, as well as examples of even more striking regularities of landholding,[34] suggest that comprehensive replanning must have been equally important. It is, after all, implied in many of the examples of adjustments to the Midland system discussed in the previous paragraph. If comprehensive replanning had occurred in the period between 1150 and 1250 the process might not have been accompanied by a written record in those townships which contained only one manor. Such a record would surely have been needed, however, in townships of divided lordship, a sufficiently numerous class to prompt questions about how they might have achieved remodelling of their field systems. Divisions and affirmations of rights in which two or more lords had an interest — rights on common moors or fens (e.g. Page, 1934, 24-7; Darby, 1940, 74-9; Hallam, 1965, 162-6; Williams, 1970, 32-8) for example, or rights of intercommoning in the fields of adjacent townships[35] — gave rise to significant numbers of written agreements in the late twelfth and thirteenth centuries. The extreme scarcity of similar records concerning agreements about divided rights in the arable which was at the core of each community's territory strongly suggests that for the most part they had already been made by the time that charters become reasonably abundant; that remodelling was, for most communities, an accomplished fact by the middle of the twelfth century.[36]

To this conclusion an important qualifying note must be added. Although, on balance and for the areas discussed here, we must reject Joan Thirsk's general conclusion that 'many villages in the Middle Ages appear to have contained numerous fields, not apparently arranged in any orderly groups',[37] the evidence of early charters does seem to indicate the presence, in places, of systems less regular and orderly than the Midland system. Without the help of ancillary sources, interpretation

of a charter which seems to reveal what I have termed an irregular
multi-field system presents many difficulties. On the one hand, J.R.
Ravensdale was absolutely correct when he concluded that 'appearances
of complexity and confusion' in the nomenclature of charters and other
sources may in many cases be a 'documentary illusion' (Ravensdale
1974, 108-9).[38] On the other hand, charters in which scribes used
formulas without mention of two or three fields may in some cases be
truthful in their indication of irregular multi-field systems. Where
twelfth- and thirteenth-century charters reveal townships with field
arrangements of this type lying in the heart of areas which both early
and later sources show to have been dominated by the Midland system,
those townships would be well worth examination in detail for possible
evidence of later remodelling on two- or three-field plans. But the
impression has been gained that most townships with irregular field
systems in counties which Gray included within his Midland zone lay
either on the fringes of that zone or in distinct pockets in its interior.
They were located in areas rich in meadowland or marsh, or near upland
areas where pasture remained abundant in the Middle Ages, and for
these reasons, perhaps, never adopted the Midland system.

Beyond the Conquest

As we move back in time beyond the 1150s, even the evidence of charters begins to fail. F.M. Stenton has shown that towards the end of the
twelfth century draftsmen of charters were sometimes still employing
'experimental formulas' which are 'interesting from their very crudity'
when they come to describe the strips of a scattered holding. Around
the middle of the century, topographical detail in descriptions of
holdings is much rarer: 'in general it may be said that the earlier the
charter the more concise will be its notes of [topographical]
identification'. In the first half of the twelfth century 'it is . . . probable
that testimony of witnesses to seisin was taken to be the best of all
evidence of the execution of a gift' (Stenton, 1920, xlvii-ix). There can
be little doubt that the gradual appearance of references to the Midland
system in the post-Conquest period is related largely to the chronology
of the documents and to the development of diplomatic. It cannot be
taken as a guide to the chronology of the system itself.

Before the Conquest the only local documentary sources which have
a bearing on the chronology of the Midland system are Anglo-Saxon
charters. These are in many senses very different from the twelfth- and
thirteenth-century charters discussed above, for most of them concern
units of land far larger than a peasant holding or a portion of demesne,

and their references to field systems tend to be incidental rather than direct. The full contribution of Anglo-Saxon charters to the history of field systems will be made only after much more detailed work has been carried out on their boundary clauses and the results interpreted in the light of local topographical information. In the meantime we must be content with their occasional explicit references to field systems.[39] For the tenth century, but not earlier, there are references in a number of charters to what are clearly intermixed acre strips. When charters refer to land lying 'acre under acre' or to 'mixed acres' they must be pointing to something very similar to the intermixed acre strips first described in detail in the twelfth century. Extensive tracts of land could be divided in this way: ten hides at Drayton (Berkshire) for example, or three hides at Hendred in the same county, the subjects of charters dated 958 and 962 respectively (Birch, 1885-93, vol. 3, 234-5, 326-7). But these references do not add up to unequivocal evidence for the Midland system, for our knowledge of later centuries shows very clearly that the presence even of large acreages of intermixed strips is no guarantee of their dispersal among two or three great fields. Only one Anglo-Saxon document is more explicit about this matter, the charter of 974 by which the Bishop of Worcester granted a lease of 30 acres 'in two fields' at an un-named place outside the bounds of Cudley (Worcestershire). 'It sounds' comments Finberg, 'as if the pattern . . . was the classic pattern of the two-field system.'[40] Also relevant are references in the charters to *gemaene land* (common land), references which may, again, relate to substantial acreages.[41] A charter for Ardington (Berkshire) contains a unique explanation of what 'common land' implied in the tenth century: 'the open pasture is common and the meadow is common and the arable is common' (Birch, 1885-93, vol. 3, 307-8). These charters, too, stop short of describing the system which we come to see in later documents. They may well relate to arrangements which had not reached the comprehensive and highly organised state which was in operation in the thirteenth century with common fallowing of whole fields every second or third year (cf. Titow, 1976). On the other hand, references as explicit as that from Ardington, or to large acreages of common arable, make it by no means improbable that a Midland system had been or was about to be adopted. So too do charters which imply that arable strips reached to the boundary of the estate, for was not the elimination of wasteland in this way a condition for the emergence of regulated common field systems? To these conclusions it is necessary to add a number of cautionary points. The crucial references are generally in the nature of incidental, 'bonus' information and their absence in any particular case can prove nothing. The great majority of them come from the southern part of the Midlands, leaving blank at least two-thirds of the area later to be dominated by the two- and three-field system, and none is earlier than the middle of the tenth century. These facts may be relevant for the chronology of the system but may also be a product of the nature of the corpus of Anglo-Saxon charters.

For earlier centuries, documentary sources provide few clues to the nature of field systems and we can do no more than assess the balance of probabilities about developments in agrarian organisation in Midland England. If we begin in the early years of the English settlement, let us say the first two and a half centuries of Saxon occupation, perhaps six points need to be made before such an assessment can be attempted. First, there is now a tendency to believe that the large, tightly-knit nucleated village, once thought to be a typical Saxon form of colonisation in Midland England, was not in fact characteristic of the earliest English settlements. 'In no case', writes Rahtz in a survey of the archaeological evidence, 'is anything like a nucleated or "green" village plan . . . discernible' (Rahtz, 1976, 58). The earliest settlements appear to have been small and loosely formed, justifying the term hamlet rather than village if we can judge from the few examples which have been excavated.[42] These were not settlements which one might expect to have been associated with extensive tracts of arable, extreme parcellation of the land or complex field systems. Second, there is now also a tendency to believe that the earliest Saxon settlers found a landscape which, in places, was not without traces and survivals of previous cultivation systems. They need no longer be envisaged as having been forced, as the Orwins' imagined pioneers were, to evolve field systems whose nature was determined by the need to survive under pioneer conditions. Third, it would nevertheless be wrong to use the accumulating evidence for pre-English occupation in the Midlands in order to argue that communities grew so rapidly and resources became so scarce in the first centuries of the Saxon settlement that settlements were forced at an early stage to adopt complex common field arrangements. The first Saxon settlements were probably unstable from the point of view of both their populations and their locations, as examples of deserted early sites show. Moreover, to what might be termed an 'ecological' stability must be added other disruptive effects on the growth of some communities: Maitland remarked long ago, and perhaps with some exaggeration, that 'readers of the English Chronicle will doubt whether there is any village in England that has not been once, or more than once, a deserted village' (Maitland, 1960 edn, 424).

The fourth and fifth points relate to the types of field system which we should perhaps envisage as belonging to these small, unstable yet uncrowded settlements of the first centuries of the English occupation. Two models are suggested by comparative evidence. In parts of the Vale of York which were extensively devastated during the campaign of 1069-70, charters allow a picture to be built up of the fields which came into being during recolonisation. In some places the land appears to have been 'cleared and cultivated by individuals' who created closes held in severalty when they reoccupied a vill.[43] Here, pioneering settlers entering a devastated but previously occupied countryside were not initially attracted by common field husbandry. The other model is suggested by recent work on what may have been the earliest field forms of the western German lands. Much diversity clearly existed and the picture is not made simpler by disagreements among German scholars. But, in addition to communities with enclosed fields held in severalty, one type of field system recurs in the literature. This is a system which seems to lie half way between separate, fenced closes and extensive subdivision into acre and half-acre strips, half way between severalty and comprehensive common rights. Each farm comprised a few large, wide strip-shaped blocks stretching back from the habitation area of a hamlet; parts of this cultivated land may have been thrown open to common pasture when not under crops, but, because of the simplicity of the field pattern, individuals could no doubt have fenced their shares if they so wished; beyond the outer perimeter of the cultivated land lay common pasture or waste.[44] Such a simple arrangement would have suited the small groups who seem to have occupied the earliest settlements in the English Midlands.

Sixth, and finally, the laws in Ine say something of the types of field arrangement which had evolved in Wessex by the late seventh century. One law maintains that the ceorl's *worthig* 'must be fenced winter and summer' so that his neighbour's cattle will not 'get in through his own gap' (translations from Whitelock, 1955, 368-9). Translations of *worthig* in this context vary. Some give simply 'homestead' (thus Whitelock), implying that this law has no direct bearing on field systems. But although the word could mean homestead, in general it was one of those Anglo-Saxon 'habitative' terms which might be applied to anything from a minute plot of land, to an urban property, to a farm, to a whole estate (Bosworth and Toller, 1898, *s.v. worth, worthig*). In the law, the reference to a 'gap' and to damage done by trespassing cattle suggests that it was being used in the sense of an enclosed farm.[45] Ine's Wessex included an active frontier for colonisation towards the west, where the element -worthy is common in the names of places which have survived as single farms to this day (Smith, 1956, vol. 2, 275-6). The law was probably directed at such farms, held in severalty from the beginning as those on the wasted frontiers in Yorkshire were to be later; it could also have applied to some farms in the heartland of the kingdom, perhaps to the largely pastoral elements in a system of specialised settlements. Another law refers to what is clearly an altogether different system in which ceorls have 'common meadow or other land divided in shares' (*gedal land*): if 'some have fenced their portion and some have not, and [if cattle] eat up their common crops or grass, those who are responsible for the gap are to go and pay to the others, who have fenced their part, compensation for the damage'. There follows a second and related passage, less often cited whose wording is not perfect, but which refers to a ceorl who finds an animal on 'his arable' (Whitelock, 1955, 368-9). These passages may be interpreted in a number of ways, for much depends upon the question: what is *gedal land*? Like the term *worthig*, it could be applied to units both very extensive and very small. In Northumbria in 876 the Scandinavians 'shared out' the land (Classen and Harmer, 1926, 29): here and elsewhere the Chronicle uses a verb with the same root as *gedal land* in the context of the partitioning of a territory among followers. At the other extreme, by the twelfth century *gedales* or *dailes* could be strips of less than an acre in size (Stenton, 1920, xlv, n. 4; Timson, 1973, 68, 222-3).[46] In Ine's law the term does not appear to have been used in either of these two senses. Charters which mention shareland provide no precise clues as to the usage of the word in the law,[47] which must therefore be interpreted in its own terms. A 'fence' or 'hedge' is crucial to the system which is being described; this is the secure boundary between cultivated land and the surrounding waste used by animals during daytime. Within the ring fence the arable core is not divided in such a way that each husbandman's land touches the boundary in many places through a scattering of strips; rather, his land is in a block so that he is responsible for a single stretch of the boundary. 'His arable' may perhaps be temporarily fenced from that of his neighbours, but in some senses the core is regarded as a 'common' shareland, perhaps because it could on occasion be used as common pasture and because of the circumstances of its creation through a process of sharing. If this interpretation of the law is correct, here was a system akin to some early continental examples and far less complex than the system of intermixed acres and compulsory fallow which was to bring forth the more detailed and intricate by-laws of a later age.

In sum, a consideration of the context of the earliest English settlements in the Midlands suggests that at least until the end of the seventh century, and perhaps long after that in many places, they were associated with small, relatively simple arable field systems. By the tenth century, more complicated systems with intermixed acre strips had developed in some townships, and it is possible that at some of these places a two- or three-field system had been or was about to be put into practice. At the end of the twelfth century there can be no doubt that the system was fully developed both in its organisation and in its extent.

Approaches to the Adoption of the Midland System

The Topographical Approach

It is clear from the discussion in the previous section that the emergence of the two- and three-field system, while by no means as early as was once believed, took place at times which are beyond the reach of documentary materials which could throw any light at all upon the details of the process or upon earlier arrangements. Once this fact is accepted it should at once stimulate alternative approaches, particularly detailed topographical work designed to assess the extent of the evidence for older and simpler systems underlying two- or three-field plans. The approach must be topographical rather than archaeological, for it is unlikely that resources will be available for the recovery, called for by D.M. Wilson in 1962, of an 'excavated Anglo-Saxon village and its field-system' (Wilson, 1962, 78).

The topographical approach, which is sometimes too grandly styled 'morphogenetic analysis' (the use of later evidence of form in order to discern earlier stages of growth) is well illustrated by the work of E.A. Pocock on the two-field village of Clanfield (Oxfordshire). From a detailed study of the size and arrangement of the furlongs within the pre-enclosure field system of Clanfield, and of differences in the amplitude of surviving headland ridges marking the ends of these furlongs, he convincingly demonstrated that prior to the emergence of a two-field system at Clanfield the arable land of the parish was created and worked by a number of different groups each with its own settlement and fields (Pocock, 1968). For Wheldrake (Yorkshire) J. Sheppard has shown, through a combined use of topographical and documentary evidence, how each of the three largest subdivided fields which existed in the township at the end of the fourteenth century had come into being over a very long period of time and was an amalgam of land units with different tenurial histories. One of the fields comprised what seem to have been the original 'toft acres' directly associated with the dwellings in the settlement at the end of the eleventh century, several furlongs which probably originated in the early twelfth century as assarts in severalty (later to be split into strips), and blocks of strips which came into being later through communal reclamation of waste. By the fourteenth century all of these units had come to be amalgamated and the whole regarded as a single field (Sheppard, 1966, 66-72). Further work of this kind is badly needed and cannot fail to throw light on the variety of types of field systems which preceded two- or three-field arrangements. Only where remodelling was exceptionally thorough, involving a radical reorientation of strips and boundaries, would we expect previous phases of development to have bequeathed nothing to the later system.[48]

The names as well as the arrangement of furlongs in Midland systems may also be regarded as important trace elements reflecting early stages of development. If the name of a furlong, now far from habitation, contains a habitative element, this suggests that it may originally have been the distinct arable land of a small community which later came to be deserted, its arable becoming absorbed into the field system of a larger and surviving settlement.[49] For a part of the West Midlands, W.J. Ford found that 'examples of field-names which suggest settlement sites . . . are widespread . . . and include names containing the elements *worth, thorp, cote* and sometimes even *tūn*' (Ford, 1976, 287). J.R. Ravensdale has given us the Cambridgeshire example of Landbeach where one of the three fields was called Dunstall Field (from *tūn-stall*, farmstead) and another Banworth Field (from *worthig*); by the early Middle Ages no settlements with these names survived either in Landbeach or in any of its adjacent parishes (Ravensdale, 1974, 101-2). Non-habitative descriptive elements are also of relevance. Elements such as *wald* or *feld* tell of stages in the expansion of a field system; the names 'Oldfield', and perhaps 'Longlands', may be indicative of an original nucleus of arable land and, where they survived embedded in two- or three-field plans, may provide clues as to the origin and growth of the systems which preceded the Midland system.[50] Personal names in furlong names must surely in most cases be evidence of an early stage of cultivation in severalty before the furlong came to be divided among a number of holdings (Bishop, 1935, 19-24; Sheppard, 1966, 68-71). Here — in the prevalence of the personal element in the nomenclature of the Midland system — is a large and very important subject for further investigation, which will be aided by the excellent collections of such names in recent county volumes published by the English Place-Name Society.

More topographical work on the arrangement of strips within furlongs is needed in order to throw light not only on older structures underlying two- or three-field plans but also on the principles which were adopted when those older structures were recast in the fashion of the Midland system. In this connection there is an urgent need for detailed studies of the disposition of glebe land, for the glebe tended to be the most stable of holdings and the most likely to take us back to early arrangements. A remarkable symmetry in the allocation of acres between the two or three fields of many Midland townships has already been noted. In some townships within the Midland zone there are hints of an even greater degree of organisation of the arable. These were townships which possessed not only a regular allocation of the acreages of holdings *between fields* but also a regular ordering of the strips of holdings *within the furlongs of each field*. Some of the best evidence for arrangements of this type is rather late and may indicate a late medieval or post-medieval reallocation of strips (Göransson, 1961, 85-98). But there is also evidence in thirteenth- and fourteenth-century sources which suggests this repetitive patterning of strips within furlongs. Many of the relevant references to the system are not easy to interpret, so that we still do not have a precise idea of its prevalence or distribution.[51] But where it occurs in early medieval sources it would seem to point to a systematic remodelling of the arable land of holdings at a time long after initial settlement. Finally, M. Harvey has recently identified even more regular structures within the two-field systems of a group of Holderness townships. Here, too, the strips of holdings were arranged in a repeated sequence but what gave a special distinctiveness to the fields of these townships was the absence of interlocking furlongs of different sizes and shapes, with their differing orientations of narrow strips, which gave an appearance of disorder to most Midland field systems; instead, each of the two fields was divided into a number of parallel units of great length, every one of these units running throughout the whole width of the field and containing the same number of wide strips arranged, from the point of view of occupancy, in the same order. Topographical analysis of this very simple and regular two-field system suggests that it was already in existence in the thirteenth century but that, being appropriate only to a fully-occupied township, it must have resulted from an act of remodelling which took place long after the first settlement (Harvey, 1978 and this volume).

The system was to be found not only in Holderness but also in some
other places in Yorkshire (Harvey, 1978, 24; Göransson, 1961, 87;
Sheppard, 1973, 149-50); it may be akin to the arrangement of
furlongs 'with strips 500-600 yards long' running 'from the scarp of
one dale to that of the next without a break' which has recently been
discovered on the ground at the deserted settlement of Wharram
Percy (*M.V.R.G.* 1978, 20). Now that its form has been fully
described, for Holderness at least, it is easier to interpret medieval
references to what may have been similar systems, in Derbyshire and
Lincolnshire for example.[52]

The Comparative Approach

An approach to the history of field systems in the Midland zone
through field forms and names which survived until a relatively late
date seems to be leading to the same conclusions as the attempt made
in an earlier section to trace them backwards in time through
documentary references and, ultimately, to speculate about their nature
at a time when all the contemporary evidence which we have is a
meagre archaeological record concerning their associated settlements.
Approached in either way the Midland system appears to have been the
result of comprehensive remodelling which came at the end of a long
period of evolution from less complex systems. But, when all has been
done and said, it will still have to be admitted that neither approach
will take us much further towards an answer to the really fundamental
questions about the nature of the forces ushering in the adoption of the
two- or three-field system which was such a major event in the agrarian
history of so many townships within the Midland zone. Here a
comparative approach may help: what clues to the circumstances of the
adoption of the Midland system are provided first by places which
never accepted it *outside* the Midland zone and second by places where
it was taken up *outside* the limits of the period which saw its general
adoption?

H.L. Gray drew too bold a boundary line around the territory of his
Midland field system. Thus, taking the example of the western boundary,
he traced it along the Devon-Dorset border, then below the eastern
slopes of the Quantocks, along the Severn shore, athwart the south-
western boundary of Gloucestershire and the western boundary of
Herefordshire, and through mid-Shropshire; to the east of this line he
saw the territory of the two- and three-field system and to its west lay
what he termed the 'Celtic system' (Gray, 1915, frontispiece). In fact
the picture was far less clear-cut than this. In the first place, the boundary

was far more diffuse than Gray imagined it to be, 'less a line than an indeterminate zone' (Treharne, 1971, 43) with some outliers to its west and, more prominent, many townships to its east for which there is no evidence of a two- or three-field system.[53] Second, intermingled in this diffuse boundary zone and between it and the field systems of the far west — the uplands of west Somerset, mid-Devon and the Welsh Marches — lay quite extensive tracts of borderland country characterised by open field systems which had a form and history different from both those of the Midland zone proper and those of the uplands. It must be admitted that Gray was not wholly unaware of some of these complexities. He considered that 'certain districts within the boundary which the thirteenth century would have drawn round the two- and three-field system seem in the sixteenth century to detach themselves' and concluded that in these districts indigenous Midland systems were 'in decay' by the time that they may first be examined in fine detail in Tudor surveys.[54]

In fact, these borderline areas for the most part had never adopted the Midland system: their sixteenth- and seventeenth-century evidence reveals not a two- or three-field system 'in decay' but the transformation of a type of field system which had never been moulded after the Midland fashion. As an example we may take the tract of east Devon countryside which lies between the western boundary of the Midland zone, drawn by Gray along the Devon-Dorset border, and the Exe some 20 miles to the west. The variegated countryside here is typical of the borderland areas, containing a mixture of extremely fertile low-lying land and patches of Greensand upland which have never yielded to the plough, except perhaps in places for sporadic outfield cultivation. In the early Middle Ages the villages of the lowlands had field systems with some similarities to those of the Midlands. Many of them were surrounded by extensive tracts of arable land; the arable lay in strips within units designated as fields and furlongs. But the diagnostic attributes of the Midland system were missing: the compact fallow field was absent and so too was the practice of village-wide common pasturing on the uncropped arable. The arable land of each settlement did not comprise two or three great fields each subdivided into furlongs, but rather a multiplicity of units — some called fields, some furlongs, some, indeed, described as both — which had not been moulded into two or three great sectors. The very earliest available evidence for East Devon, from the early thirteenth century, reveals the presence of these irregular multi-field systems and no hint of more regular two- or three-field arrangements. By the sixteenth century most of the arable had been enclosed. The agrarian history of the region is not one of the decay of a two- or three-field system but of the transformation of an indigenous and flexible multi-field arrangement (Fox, 1972).

The same type of multi-field system was to be found in the Middle Ages almost everywhere beyond the borders of the Midland belt: in much of Herefordshire (Roderick, 1951), for example, or in the variegated countrysides on the fringes of the southern end of the Pennine chain in Derbyshire and Nottinghamshire.[55] It was identical to the type of system which in the heart of the Midlands was probably the immediate prototype of fully-fledged Midland systems. Why did it fail to develop further? The answer seems to lie in the nature of the settings in which it was found, countrysides which had access to pasture in quantities far greater than those of the Midland heartlands. When, in 1365, the reeve of the borderland manor of Knapp Fee (Somersetshire) explained that he had no profits from pasture rents from part of the demesne 'because of the great abundance of pasture in the moors' or when a jury from the East Devon manor of Aylesbeare explained before the escheator in 1377 that 180 acres of rough pasture were worth only 2s to the lord 'and no more because the tenants of the manor and other neighbours pasture there', they were both making the same point: in these borderland regions pasture was everywhere abundant beyond the bounds of the arable.[56] Taking East Devon again as an example, we find that most manors possessed some meadowland in the thirteenth century and early fourteenth and also small acreages of permanent pasture, said in some extents to be reserved for 'oxen' or for 'working oxen'. But by far the greatest acreages of pasture were on the tops of the hill ranges which run between the East Devon valleys; some woodlands on the hill slopes were also used as pastures. Almost all of the larger manors of the region contained some rough grazings, described variously in extents as *mora, vastum, bruera, terra montanea* or simply *mons*. The hill ranges yielded turf and gorse and might also in places be sporadically cultivated, but their principal function was as grazing grounds, especially for sheep. Some were vast, Uplyme's hill grazings for example supporting 700 sheep from the demesne alone; their abundance was reflected by an absence of stinting and of firmly drawn boundaries between the moor of one manor and that of the next (Fox, 1972, 97-100). Under these circumstances of abundance of upland pasture adjacent to almost every manor, there was no need for the grazing which took place on the arable to be rigorously and minutely organised.

Comparison with the situation which prevailed within the Midland zone in the thirteenth and early fourteenth centuries reveals a fundamental contrast in provision of pasturage. Here early medieval extents of manor after manor make no reference to pasture other than meadowland; a few explicitly state that there is no pasture; often the only potential grazing grounds outside the meadows and arable were woods or parks but in extents these were often written off as pastures because of excessive shade cast by trees, because of possible damage to underwood or because of the presence of deer.[57] The contrast is nowhere more accurately indicated than in valuations of pasture from the relatively few manors in the heart of the Midlands and from the many manors in borderland areas which contained this type of land use. Values varied, of course, according to the type and usage of pasture, but if we take only one type, the several pastures, it may be assumed that variations in valuation reflected their local abundance or scarcity. The mean value per acre of several pasture from extents of East Devon manors drawn up between 1244 and 1347 was 1.8d per acre; Leicestershire extents drawn up during roughly the same period yield a value of 12d per acre.[58]

What is clear from this comparison between areas bordering the Midland zone on the one hand and the interior of the zone on the other is that a very real contrast in their provision with pasture at the end of the early Middle Ages may plausibly be related to differences in their field systems. Certainly, presence of pasture in abundance would not have tended to encourage transformation of the irregular multi-field systems of the borderlands into a type of arrangement whose rationale was a regular and systematic alternation of crops and fallow grazing. It is only a short step to suggest that in the Midland heartlands a decline in the availability of pasture was the stimulus which had led to the adoption of the two- and three-field system.

The comparative approach may be extended not only to regions outside the Midland zone but also to places which took up the system outside the limits of the period which saw its general and widespread adoption. in an earlier section of this paper it was shown that this period, whenever it may have begun, was certainly over by the middle of the twelfth century and that all of the documents which are so often cited as illustrating post-Conquest examples of remodelling of fields in fact concern developments other than the arrival of the Midland system. Two cases remain to be discussed, the events at Dry Drayton (Cambridgeshire) in the 1150s and at Segenhoe (Bedfordshire) in the 1160s. Neither has been evaluated in the context of the debate over the origin of fully-fledged field systems in the Midlands, yet both seem to be relevant to it.[59]

A transcript of part of the crucial evidence from Dry Drayton has been in print for a number of years. Taken from Wrest Park Cartulary, it was included in the appendix of documents which F.M. Page added to her *Estates of Crowland Abbey*, though she made only slight mention of it in her main text, in order to illustrate the types of agreements which were sometimes necessary in vills with divided lordship (Page, 1934, 19, 162). The evidence, however, is of interest not only as an example of intermanorial affairs. The instrument transcribed by Miss Page is headed *Diuidenda inter Eduardum abbatem nostrum et alios dominos de Draytone super noua partitione terre in Draitona*. It briefly recounts how the five lords of the vill agreed on a 'new partition' of 'the land of the vill both field land and meadow'. The reason for this 'exceedingly necessary' action was the fact that the land had been 'for long uncultivated' as a result of being 'dispersed in minute parts'. The new system was to be marked out 'by certain boundaries' and *per amplas culturas*. The printed transcript tells us no more but a note added by the cartulary's scribe, and not copied by Miss Page, helps in the dating of the instrument to the 1150s. Three other documents follow in the cartulary. One is a royal confirmation of the partition, by Henry II. The other two refer back to the partition, describing it as a *commutacio;* one of them concerns an adjustment of six acres to the original plan.[60]

This very unusual series of documents, the bare and formal instruments ratifying what must have been a highly complex operation, still leave us uncertain as to the precise nature of the agreement which was reached by the five lords of Dry Drayton one season in the 1150s, or of what they planned together on the ground itself. But this much may be surmised. The agreement concerned the whole cultivated land of the vill and was not merely one of those relatively frequently recorded early medieval compositions about division of inter-commoned wasteland. It touched deeply enough on the whole agrarian regime of Dry Drayton for the vill's lords to think fit to secure a royal confirmation. The new arrangement was thought to be necessary because the land was 'dispersed in minute parts'. This may mean that the lands of each manor in the vill were inconveniently intermixed, but a more likely explanation, given the statement that the old system had resulted in a decline of cultivation, is that there was a more fundamental flaw in the old field system of Dry Drayton, perhaps a lack of adequate arrangements for grazing in order to keep the land in heart: strips may still be cultivated if they are inconveniently located but not if they are inadequately manured.[61] The new arrangements involved division (*dividenda*), partition (*partitio*) and exchange (*commutacio*). They were to leave their mark *per amplas culturas*, indicating that some kind of physical rearrangement of boundaries took place within the field system.[62] It seems almost certain that these sparse transcripts in the Wrest Park Cartulary represent the formal records relating to the arrival there of the three-field system which was certainly being used in the vill in 1298 (Postgate, 1964, appendix 1).

The documentary material from Segenhoe is even more intriguing and unusual than that from Dry Drayton. It takes the form of a narrative interspersed in an extent of the lands of Dunstable Priory at Segenhoe. The extent, written in the early thirteenth century, is so randomly rubricated and interspersed with narrative as to seem like a puzzle to those used to reading the more soberly composed surveys drawn up under the influence of models such as the *Extenta Manerii*.[63] Luckily, it ends with a brief summary of the field system which is being described, a two-field system in which the land of Dunstable Priory comprised 38 acres in one field and 42 in the other. The ultimate sources of the narrative, vivid and intricate yet recording events which occurred almost a century before it was written down, are uncertain.[64] It begins by recounting the details of a division of the manor between two lords, probably at some time during the first three decades of the twelfth century. One took for himself the whole park and demesne and, although the narrative is not at its clearest at this point, it seems that the other was compensated by being given a disproportionate share of the arable and woodland, including a wood 'then called Cherelewood now Norwood' which was in the process of being opened out by a small settlement of *rustici*. Thus far the narrative has touched only upon a manorial division, although the details which it contains give some idea of agrarian conditions at Segenhoe in the reign of Henry I: the vicinity was well wooded and contained a number of settlements, some of which were small clusters of peasant dwellings associated with woodland assarts.[65] The narrative then moves to the 1160s. By this time not only the earlier manorial division, which seems to have given both lords intermingled shares, but also unjust seizures of land 'at the time of the war' (presumably in Stephen's reign) and, no doubt, continued assarting had all produced complexities which made necessary a reorganisation of the field systems. Accordingly, at the court of each lord and by the knowledge of six old men, 'knights, free men and others . . . surrendered their lands under the supervision of the old men and by the measure the perch, to be divided as if they were newly won land, assigning to each a reasonable share'. Here, with some further details about how the redistribution was made, the narrative comes to an end. Its purpose in the context of the extent was clearly to explain the provenance, through all these changes, of the lands which Dunstable Priory began to acquire at Segenhoe in the twelfth century. The narrative enticingly takes us a little further towards the processes of remodelling than do the more abstract documents from Dry Drayton, although it does not state what type of field system was constructed at Segenhoe in the 1160s. There can be little doubt, though, that it was the two-field system which was mentioned there in the early thirteenth century, a reconstruction of the irregular collection of assarted furlongs which can be glimpsed in the reign of Henry I.

Dry Drayton and Segenhoe have much in common. They were both vills of divided lordship, a circumstance which may in part explain their tardiness in remodelling their field systems and which certainly has much to do with the survival of documentation relating to this process. Further, both vills were situated in parts of the belt of Midland field systems which retained some uncultivated land until a relatively late date. Dry Drayton lies only just above the 100 foot contour but was none the less dry in comparison with the other Cambridgeshire Drayton, Fen Drayton near the fenland edge. Its alternative name was Wold Drayton, under which guise it appears in one of the documents relating to the remodelling of its field system. The map shows that it lies on the flank of a minor upland area which has recently been identified as a 'wold' or former woodland (Everitt, 1977, 11); the names of its neighbouring settlements of Lolworth, Boxworth, Elsworth, Childerley, Madingley and Hardwick testify to relatively late colonisation; the district still retained some woodland in 1086, although, significantly, only in the small quantities which the Domesday scribes dismissed as 'wood for the houses' or 'for making fences' (Darby, 1952, 297-300). The name Drayton itself seems usually to have had a connection with woodland exploitation (Everitt, 1979, 70). Segenhoe is found to be in countryside with characteristics similar to those of the west Cambridgeshire wolds. In 1086 there was woodland for 300 swine as well as ten rams per year from customary woodland dues (*V.C.H. Bedfordshire*, vol. 1; 251): this was the land which we can see under attack in the early twelfth century. The manor lay between the Chilterns and the clay vales in the heart of the Midlands. The vicinity was still relatively well wooded in 1086 (Darby and Campbell, 1962, 32), nearby places are named Eversholt and Crawley, while the element *hōh* of Segenhoe itself seems in many cases to have had an association with woodland (Everitt, 1979, 70).

The reason for the very late adoption of the Midland system at Dry Drayton and at Segenhoe should now be clear. Both are situated in countrysides which came to be dominated by two- or three-field systems (Postgate, 1964, appendix 1; Roden, 1973, 345) and which by the early Middle Ages stood in marked contrast to the lands outside the borders of the Midland belt, with their abundant pastures. Yet within the Midland belt, and even if we except its uplands such as Charnwood or Arden 'Forests', there were subtle regional differences in the courses taken by settlement and colonisation. Some regions, those characterised as wolds for example, came to be well exploited later in the pre-Conquest period than did the lower-lying areas. They therefore experienced a shortage of pasture at a relatively late date. The lesson to be learned from Dry Drayton and Segenhoe is not that they represent accidentally documented examples of a common development in the centuries after the Conquest but that they are illustrative of a minority, the very last places to adopt the two- or three-field system within the Midland belt. These two places bring us back once again to the essential features of the Midland system, determined as they were by the need for systematic grazing of the arable. They also lend much support to the suggestions of Joan Thirsk about the relative chronology (but not the absolute dating) of events leading up to the maturation of field systems.

Field Systems and Settlement Systems: a Suggestion

It remains to ask about the possible circumstances under which shortage of pasture began to be experienced in the heart of the Midlands, circumstances which brought to bear upon communities and lords the need to introduce the arrangement for systematically integrating grazing and arable which was the Midland system. Pressure of population encouraging attacks by the plough on reserves of pasture is, of course, a recurrent theme in writings about the twelfth and thirteenth centuries. It would have been convenient indeed if this theme could have been linked to the development of mature field systems in the Midland belt. But the evidence on the chronology of the system, presented earlier in this paper, is altogether against such an interpretation. Moreover, it is becoming increasingly clear that in many parts of the Midlands reserves of pasture had become scarce well before the Conquest. From the viewpoint of the manorial documentation relating to reclamations, the assarts of the twelfth and thirteenth centuries represent 'the concluding state of a movement long past its climacteric', deeply affecting only the more marginal lands; from the viewpoint of Domesday Book many parts of the Midlands appear to have been 'an old country' which had 'passed beyond the colonial stage' by 1086; from the viewpoint of Anglo-Saxon charters it has been claimed that in places 'the rural resources of England were almost as fully exploited in the seventh and eighth centuries as they were in the eleventh (Postan, 1966, 550; Lennard, 1959, 1, 3; Sawyer, 1974, 108). It would have been convenient, too, if the emergence of the Midland system could have been linked to another major theme in the agrarian history of the twelfth and thirteenth centuries, the replacement on many estates of a policy of farming out manors by the direct exploitation of demesnes. Might not remodelling of field systems have accompanied the resumption of demesne farming, producing a degree of uniformity of organisation at an estate level where previously the differing policies and activities of different lessees had prevailed? But again there is no correspondence of chronologies. We can now point to the last decade of the twelfth century and the first few decades of the thirteenth as having witnessed a major shift towards direct exploitation of demesnes (Miller, 1971, 10-12; Harvey, 1974), yet the Midland system had developed some time before this.

Instead it is to far earlier, pre-Conquest trends in resource exploitation and estate organisation that we must look for circumstances conducive to the adoption of the Midland system. Here we come to topics which are dark indeed, recently illuminated here and there only by the studies of T.H. Aston, A. Everitt, H.P.R. Finberg, W.J. Ford, M. Gelling, G.R.J. Jones, C.V. Phythian-Adams and P.H. Sawyer. But what emerges from much of the recent work on these problems is the importance in the early Saxon period of the multiple estate – and of links within it between pastoral and arable settlements – as well as the importance later of processes leading to its fragmentation and the fission of its individual elements.

In terms of settlement the multiple estate was essentially a federation of township units; in terms of organisation it was centred upon a *caput* at which renders in kind were received from dependent townships. The townships differed one from another in the character of their renders. The most complete evidence for the nature of this differentiation of function within estate units comes from parts of Britain far removed from the Midlands, for it was in these areas that the system survived longest. But within the Midlands it left its mark on numerous place-names of the type Cheswick, Hardwick and Shipton, or Ryton and Barton (respectively indicating pastoral and arable specialisations). Movements within a system of this kind were not all in one direction – the rendering of *feorm* from dependencies to *caput* – for there were linkages, too, *between* dependent vills, taking the form of movements of livestock (transhumance is perhaps in some cases too strong a term) between 'river and wold', between felden and woodland. However old the system may have been, it was subject to disruptive forces. To T.H. Aston it is 'the distintegration of the primitive estate' through 'sales and mortgages, leases and rewards, inheritance and family arrangements, forfeiture, piety, illegality' which gives vitality to the history of land units as seen through Anglo-Saxon charters. To P.H. Sawyer, 'the main development in the settlement history of Anglo-Saxon England was . . . the fragmentation of large multiple estates'. To G.R.J. Jones, 'fission appears to have prevailed over fusion' in the history of the multiple estate and was 'more advanced in the richer areas' in the centuries leading up to the making of the Domesday Survey. To A. Everitt, places in areas with the characteristics of wolds began to emerge as independent settlements 'predominantly in the middle- or later-Saxon period' (Aston, 1958, 77; Sawyer, 1974, 108; Jones, 1976, 38; Everitt, 1977, 18).

An extremely complex system of settlement organisation and a regionally diverse history of its disintegration have been summarised here in far too simple a fashion. Yet the implications of both for the development of field systems are very great indeed. The multiple estate, it can be suggested, was no seedbed for the development of a type of field system whose distinguishing feature was a rigorous integration of arable and common pasture. Vills which were predominantly pastoral in their specialisation no doubt grew some arable crops for their own subsistence needs, but it is difficult to envisage situations in which they were forced to use the whole of their land alternately for grain and grazing for this would have threatened the very basis of their specialisation. Vills which were largely concerned with the production of crops might well have needed to organise their arable land on a basis more intensive than the three-field system; through reciprocal arrangements with pastoral settlements which ensured that they benefitted from the wintering of livestock as well as the removal of livestock in the summer months, they could afford to do so. If we now consider the implications for specialised and linked vills of their severance from the framework of a multiple estate, it is clear that this could only have had a fundamental impact on their field systems. Pastoral and arable vills alike would have begun to feel the turn of the screw, the more so if they passed out of the framework of a large estate to become independent units, the endowment of a knight, for example. For the former, the principal requirement would perhaps have been an increase in arable acreage at the expense of pasture; for the latter, the regular provision of grazing on already extensive arable lands. For both, adoption of the Midland system might eventually be the appropriate solution. It could be argued that only where vills had access to very plentiful supplies of pasture *within* their boundaries (as was the case beyond the limits of the Midland zone) could such a solution be delayed or avoided.[66]

The attractiveness of this model which links the arrival of the Midland system to changes taking place in a system of settlements lies in the concordance of the chronologies of the two developments: both were features of the agrarian history of the centuries immediately preceding the Conquest. In addition, it is in accordance with what is known for certain about the essential features of the Midland system as a close and closely worked out integration of arable and pastoral functions. Further, it may help to explain the rapidly increasing body of evidence which is coming forward for a trend towards the fusion and nucleation of settlements in the late Anglo-Saxon period. It must be admitted that to put such a model to the test will never be an easy exercise. For many areas in the Midlands all that we know is that settlements once with such different functions as the Weeks or Swinnertons and the Bartons, with such differing status as the Kinghams and the Knightons, with such different pedigrees as the Walcots and Worthys and the Thorpes,[67] had all by the thirteenth century come under the rule of the Midland system; their early tenurial history in relation to nearby places, let alone details of the evolution of their field systems, may always remain unknown. Perhaps it will be on the borders of the Midland belt — where early estates straddled both 'Midland' and 'borderland' countrysides, and where there was divergence rather than convergence in the development of field systems in different settlements — that investigations will best be directed. There, with the help of topographical evidence, and in particular with the aid of pre-Conquest charters which can be made to yield up hints about agrarian details as well as clues to the organisation and disintegration of early estates,[68] may perhaps be found some degree of confirmation for the model which has been proposed above, as a working hypothesis rather than an article of faith.

Whatever the final verdict may be, it should be clear that the 'advent and triumph' of the Midland system are to be sought not in the high Middle Ages but in the later centuries of the Saxon period. The movement may well have been at its height at a time when Viking influence on the course of settlement history was at its strongest, though this is not to say that the system was a Scandinavian introduction:[69]

rather, the expansion of old settlements in the tenth century and the establishment of new communities (on lands hitherto reserved as pastures?) should perhaps be seen as catalysts which precipitated its adoption. To these generalisations on chronology there may well be a few exceptions: in parts of the North devastated during the campaign of 1069-70 a two- or three-field system may have developed after the Conquest;[70] elsewhere there are perhaps examples yet to be discovered of that minority of places which followed the same course as Dry Drayton and Segenhoe. But for the most part the origins of the system were far earlier and must be sought not through predominantly documentary studies as admirably exemplified in *Studies of Field Systems in the British Isles* but through an approach which is concerned far more with the early history of settlements and estates and which makes full use of topographic and toponymic as well as of documentary evidence. Undoubtedly the finer details of the diffusion of the Midland system will always remain obscure. Indeed, we may never know whether or not its development in England really was a true diffusion process. On the one hand there are examples of places which seem to have taken up the system imitatively without having experienced the stimuli and pressures which have been stressed throughout this paper.[71] On the other hand there is much to be said for the notion that communities arrived independently at the same solutions as they approached towards the adoption of the Midland system.

Appendix: *Evidences of the Midland system before circa 1250: Gloucestershire, Cambridgeshire, Lincolnshire*

Gloucestershire

Abbreviations. C: *The Cartulary of Cirencester Abbey* (Ross, 1964; Devine, 1977) P: *Historia et Cartularium Monasterii Sancti Petri Gloucestriae* (Hart, 1863-7) W: *Landboc sive Registrum Monasterii Beatae Mariae de Winchelcumba* (Royce, 1892-1903). Other sources cited refer to items for which full details are given in the bibliography. Adlestrop (*V.C.H. Gloucs.* VI, 13); Alderton (P, I, 167); Aston Blank (W, I, 233-4); Aston Subedge (Salter, 1906-8, I, 137-8); Badminton (Gray, 1915, 464); Baunton (C, 228-9); Bishop's Cleeve (Hollings, 1934-50, 351); Cherington (*V.C.H. Gloucs.* XI, 171); Cirencester (C, 259); Cowley (Gray, 1915, 464); Culkerton (*V.C.H. Gloucs.* XI, 241); Cutsdean (W, II, 304-6); Duntisbourne Abbots (C, 1000-1); Eastleach (P, I, 271-2); Frampton Mansell (C, 351-3); Hampen (W, I, 151-2);

Hawkesbury (Gray, 1915, 465); Hidcote (Salter, 1906-8, I, 137-8); Leighterton (P, I, 358-9); Little Rissington (*V.C.H. Gloucs.* VI, 110); Mickleton (Salter, 1906-8, I, 137-8); Nutbeam (C, 268-9); Rodmarton (C, 344-6); Sherborne (W, II, 234-5 *et seq.*); Shipton (Gray, 1915, 465); Stratton (C, 277-8); Througham (C, 945); Whittington (Gray, 1915, 465); Yanworth (W, II, 320-1)

Cambridgeshire (including the Isle of Ely)

Abbreviations. P: Appendix 1 of Postgate (1964) S: Survey of the estate of Ely, Ely Diocesan Records (Cambridge University Library), G. 3. 27.
Abington Piggots (P); Bassingbourne (P); Cheveley (P); Cottenham (P); Downham (S f. 7v.); Eltisley (P); Eversden (*V.C.H. Cambs.* V, 63); Great Abington (P); Great Wilbraham (P); Haddenham (P); Linden (S f. 19v.-20); Linton (P); Littleport (S f. 10v.); Little Wilbraham (P); Rampton (P); Shudy Camps (P); Tadlow (P); Toft (*V.C.H. Cambs.* V, 132); Triplow (S f. 72); Waterbeach (Ravensdale, 1974, 88); Wilburton (S f. 17-17v.); Willingham (S f. 58)

Lincolnshire

Abbreviations. D: *Documents Illustrative of the Social and Economic History of the Danelaw* (Stenton, 1920) G: *Transcripts of Charters Relating to . . . Gilbertine houses* (Stenton, 1922) R: *Registrum Antiquissimum of the Cathedral Church of Lincoln* (Lincoln Record Society, 10 vols, 1931-73). Other sources cited refer to items for which full details are given in the bibliography.
Alvingham (R, V, 99-100); Asgarby (R, II, 192-3); Aylesby (Brown, 1889-94, I, 315); Benniworth (R, V, 55-6); Binbrook (R, IV, 233-4); Brocklesby (D, 186-7); Burgh on Bain (R, V, 3-4); Burton by Lincoln (R, IV, 83-4); Cadbourne (R, IV, 262-3); Cadeby (G, 28); Carlton-le-Moorland (R, VII, 55-7); Claxby (R, IV, 204-5); Claxby Pluckacre (R, VI, 100-1); Corringham (R, IV, 4); Cotes (G, 83); Croxby (G, 54); Dunholme (Stenton, 1969, 137); E. Barkwith (G. 94); Edlington (D, 128-9); Fillingham (R, IV, 94-5); Fotherby (G, 51); Friesthorpe (R, IV, 112-3); Habrough (D, 199-200); Hackthorn (D, 22); Hainton (G, 10); Hameringham (R, VI, 141-2); Harlaxton (Foster, 1920, 62); Heckington (R, VII, 113-5); Hibaldstow (R, IV, 36-7); Holton by Beckering (R, V, 6-7); Ingham (D, 41-2); Laceby (R, IV, 269-70); Linwood (R, IV, 222-3); Lissington (R, V, 63-4); Middle Carlton, (R, II, 251-2); Nettleton (R, IV, 145-6); Newton by Toft (R, IV, 169-70); Normanby le Wold (R, IV, 190-1); N. and S. Kelsey (R, IV, 139-40); N. Reston

(R, VI, 51); Norton (Massingberd, 1896, 284); Owersby (R, IV, 161-2);
Owmby (R, III, 367-8); Oxcombe (G, 97); Rand (R, V, 39-41);
Redbourne (D, 60); Saltfleetby (R, V, 123); Scamblesby (R, VI, 156-7);
Scawby (R, IV, 81-2); Scopwick (D, 324-5); Snelland (D, 155-6);
Somersby (R, VI, 138); S. Cadeby (R, V, 76-7); Stixwould (R, VI,
160-1); Stubton (Massingberd, 1896, 244-5); Swaby (R, VI, 39);
Tetford (R, VI, 123-4); Thorpe le Vale (R, V, 44); Thurlby by Lincoln
(R, VII, 180-2); Welbourn (R, VII, 149-50); W. Ashby (R, VI, 146-7);
Withcall (R, V, 82); Wyham (R, II, 284-5); Yarborough (R, V, 117)

Notes

1. This paper was originally prepared to form the basis for discussion at the
Oxford conference on 'The origins of open-field agriculture'. Many of the topics
which are touched upon need far deeper treatment than I have been able to give
them here; I hope to expand upon some of them in amended and enlarged versions
of parts of the paper, to appear later. I would like to record the debt that I owe
to Sandra Raban of Homerton College, Cambridge. Not only did she bring to my
attention, in 1968, a remarkable document from Dry Drayton (discussed later in
this paper) but she also assisted me, ten years later, by providing transcripts of
other related materials. I am also grateful to Mr C.V. Phythian-Adams for giving
me access to his knowledge of the field reviewed here.

2. Work since 1915 had led to some slight revisions of the course of the
boundary of the Midland system as drawn by Gray but, with the important
exception of parts of Northumberland (Butlin, 1964, 101-5), the broad outlines
of the distribution which he described seem to be substantially correct.

3. Even where the land lay enclosed and in severalty, as it did in most parts
of Devon and Cornwall after about 1500, the basic fabric of the field system – its
hedges, lanes and other boundaries – was regarded as a proper area for common
concern and common overseeing: Fox (1977), 59.

4. *Et peccavi:* Fox (1975), 202.

5. This is quite clear from extents which describe numerous furlongs and
fields as well as a system of two or three seasons. Examples are extents of the
Bishop of Chichester's manors of Sidlesham, Greatham and Aldingbourne, of St
Albans Abbey's manors of Codicote and Tyttenhanger and of the St Paul's manor
of Navestock: Peckham (ed.) (1925), 127-9; Levett (1938), 183, 338-9; Hale
(ed.) (1858), cxxii. Pelham (1937), 197-8 was mistaken when he inferred a three-
field system from these Sussex extents. For a clarification of Miss Levett's
interpretation, see R. Lennard's editorial note (in Levett, 1938, 183, n.2) and
Roden (1969), 22, n.1. Codicote and Tyttenhanger were townships in the
Chiltern Hills where the most usual arrangement in the thirteenth century was for
a season to comprise 'fields lying throughout the common arable and not
necessarily adjacent to each other': Roden (1973), 337.

6. In medieval extents the term *seisona* could on occasion be equated with a
whole *campus,* as in a thirteenth-century extent of Louth (Lincolnshire): Queen's
College Oxford, MS. 366. But, in general, the term was most commonly used in
extents of multi-field manors where a word other than *campus* was needed to
describe the basic unit of rotation. This was the case on the St Paul's estate where
the articles of visitation of c. 1290 stipulated *qualibet seisone distinguuntur* as

well as asking *quot campi sunt in dominio?*: Hale (ed.) (1858), cxxii. Those who drew up an extent of Littleton (Hampshire) in 1265-6 seem to have been unaware of the precision which could be obtained by employing both terms. After listing a variety of 'furlongs' and *campi* which made up one unit of rotation, they added, by way of summary, the note that together they all *jacent pro uno campo;* here the term *seisona* would have been more appropriate. The extent is printed in Hart (ed.) (1863-7), vol. 3, 35.

    7. From the point of view of cropping a system where the season comprised a number of detached blocks was neither more nor less convenient than a system where the season comprised a single great field within which the land of each holding was widely scattered. But from the point of view of fallow common to all animals of all villagers the latter was more desirable than the former simply because it did away with the inconvenience of allocating stock to particular units or of frequently moving them from one unit to another. It was desirable, too, because it removed stock to one sector of a village's land, thus minimising the risk of their trespassing on growing crops.

    8. There had always been some doubt about this attribution of the two- or three-field system to the material culture of the Anglo-Saxons. See, for example, Seebohm (1883; 1905 edn), 410-11 and Barger (1938).

    9. I give some of the evidence for this view in a subsequent section.

    10. Titow (1965), 94 and *passim.* Titow does, however, stress that his aim was not to defend 'the orthodox view about the open-field system in medieval England' and he agrees that mature field systems 'might have developed some time after' the arrival of 'bands of immigrants . . . faced with the necessity of carrying out some distribution of land from the very start': ibid., 102, 94. See also Titow (1969), 19-23.

    11. British Library, Eg. MS. 3321 and, for Walton, Longleat House Muniments, 10024 transcribed by Keil, 1964, appendix 2; these extents are more detailed and consistent than those printed in *Rentalia et Custûmaria,* 1891. The Glastonbury estate also included manors in areas not characterised by the Midland system, the Deverill manors of the Wiltshire chalklands, for example. I am currently making more detailed study of the field systems portrayed in these surveys, their management in the fourteenth century and their origins.

    12. Ely Diocesan Records (Cambridge University Library), G.3.27; Staffordshire County Record Office, D(W) 1734/J 2268. Both estates likewise included manors which were not organised on the lines of the Midland system, for example the fen-edge and Suffolk properties of the Bishop of Ely, for which see Miller (1951), 80, n.1.

    13. For the Fortibus estate see Gray (1915), 504-6, citing P.R.O. S.C. 12 17/4. For the estate of the Bishop of Worcester, extents in the Red Book (Hollings (ed.), 1934-50) vary in the quality of their descriptions of demesne arable. Some (e.g. Henbury in the Salt Marsh) seem to describe irregular multi-field systems in which a two-course rotation prevailed. Others (e.g. Hampton Lucy, Old Stratford) describe a two-field system without specifying the acreages of the fields. Finally, there are two extents which describe, without doubt or ambiguity, near perfect two-field systems (Withington and Bibury, Gloucestershire). The terminology of extents in the thirteenth-century survey of the estate of the Bishop of Lincoln (Queen's College Oxford, MS. 366) is also frustrating, but the estate could claim several probable examples of well developed Midland systems and one case in which there is no doubt (Louth, Lincolnshire).

    14. The degree of dispersal of demesne lands within the fields varied considerably and no doubt depended partly upon the nature of the origin of the demesne, partly upon whether or not lords had thought fit to carry out policies of consolidation. There was a range of types, from strips dispersed among the

furlongs, to compact furlongs, to blocks of adjacent furlongs in each field. This range is illustrated by examples in Hilton (1947), 54-5; Harvey (1965), 20-2; and Stenton (ed.) (1920), 140-1. It is rare for the documents to state explicitly that demesnes were or were not intermixed with the lands of tenants, but occasionally they do so: Raine (ed.) (1864-5), vol. 2, 72; Willis (ed.) (1916), 74. Some degree of intermixture (or, at the very least, demesne lands which marched with the village lands in the fallow season) is surely implied by frequent references in extents to demesnes of which a specified proportion each year lay 'fallow and in common'. There are many examples in Gray (1915), 450-509.

    15. British Library, Harl. MS. 3961 and Eg. MS. 3134 and 3034.

    16. Even the printed examples are literally too numerous to cite. A representative sample of abstracts from charters illustrative of the two- and three-field system is given in Gray (1915), 450-509. Some of the earliest relevant charter evidence from two parts of the country is listed in an appendix to this paper.

    17. For example, Rees (1924), 192-3 (S. Wales); Gras and Gras (1930), 33-8 (Hampshire Downs); Bishop (1938) (Kent); Keil (1964), appendix 3 (Longbridge and Monkton Deverill, Wiltshire chalklands); Brandon (1962), 65-8 (South Downs); Britnell (1966), 383-5 (Essex); Roden (1966), 50-1 and (1973), 337 (Chiltern Hills).

    18. Harvey (1965), 164-5; Roderick (1951), 57; Keil (1964), appendix 3; Keil (1966), 238-9; Roden (1973), 349; Ravensdale (1974), 106-7; Page (1934), 119; Hilton (1947), 153-5. The evidence from Oakington appears to show what Miss Page described as 'experiments . . . in dispensing with the fallow field' for a short period between 1383 and 1391.

    19. What, for example, is to be made of explicit references to 'the sown field' or 'the fallow field' if they do not imply a regular alternation of land use between whole fields? At Piddington (Oxfordshire) we hear in *c.* 1106 of a system in which either the west field or the east field was sown in any year, surely one of the earliest explicit references to a two-field system if the dating of the source is correct; an extent of South Stoke in the same county says of the three fields that 'two are sown each year and the third is fallowed'; another Oxfordshire extent, of Kidlington, declares that one-third of the arable 'lies fallow and in the common field'; at Cottenham (Cambridgeshire) accounts use the term 'falufeld'; at Harlestone (Northamptonshire) the agreement of 1410 states that 'each year one field . . . is sown with wheat and barley and another field with beans and peas and the third field lies for fallow'; at Bishop's Hampton (Warwickshire) it was ordained in 1460 that 'one field lie fallow in turn every year': Gray (1915), 488; Salter (ed.) (1906-8), vol. 2, 120; Gray (1915), 491; Page (1934), 119; Wake (1922), 410; Ault (1965), 33. A grant of *c.* 1150 relating to the two-field vill of Newhouse (Lincolnshire) states that half of the land shall be cultivated each year; the more abbreviated terminology of a charter of roughly the same date from Thorpe le Vale may imply the same; a lease of a holding in Hackthorn in 1238 stipulates that the lessee shall take five crops from the north field and five from the south field; there is a similar stipulation in a lease from Binbrook in 1240: Stenton (ed.) (1920), 232-3; *Registrum Antiquissimum*, vol. 5, 44; vol. 4, 69-71; vol. 4, 233-4. For another ten-year lease (relating to Culkerton, Gloucestershire) with the same provision as that for Hackthorn, see *Transactions of the Bristol and Gloucester Archaeological Society,* 22 (1899), 204-5. Agreements by which two adjacent vills made arrangements for the inter-commoning of their fallow fields (see below, n.35) imply more regularity in the working of the system than some historians would allow. It could be argued too that use of a special term − 'inhok' − to describe a portion of the fallow which was used for cropping, as a short-term expedient rather than as a long-term modification to the field system, suggests that the concept of a fallow field was a very firmly rooted one. For some

examples of the practice see Gray (1915), 92-3; Homans (1941), 57-8; Hallam (1972), 219; Thirsk (1973), 261.

20. This paragraph relies heavily upon the earliest references given in Ault (1965).

21. The 'inhok' (for which see above, n.19) was a common means of achieving flexibility in cropping within the Midland system. For the ways in which peasant assarts could be incorporated into an existing field system see the excellent description in Bishop (1935), 18-19. An interesting example of how an increase in demesne acreage (perhaps through reclamations) could be accommodated in a three-field system is provided by a note appended to the 1251 extent of the Bishop of Ely's manor of Wilburton: Ely Diocesan Records (Cambridge University Library), G.3.27, f. 17. In 1251 there were three fields, two with an approximately equal acreage of demesne, the third with a slightly smaller acreage. A later measurement, added at the foot of the 1251 extent, shows how two of the fields had been expanded and the third slightly reduced in acreage. As for exchanges, there is plenty of scope for what would undoubtedly be a revealing study of the motives for the transfers of strips so frequently encountered in medieval sources relating to two- or three-field systems.

22. Twelfth-century surveys do, however, contain some explicit references to the Midland system in formulae whose purpose seems usually to have been to explain the atypical nature of a minority of tenant holdings with land unequally distributed among the fields. The evidence is neatly summarised in Titow (1965), 98, n.7-10.

23. Ross (ed.) (1964), 343-6. Only slightly more useful are charters which refer to land *in territorio de, in campo de* or *in campis de* x (the name of the place). There is a large element of common form in the use of these terms although they do generally seem to have been applied to holdings comprising scattered lands rather than to compact blocks. There is usually little point in attempting to find a significance in the use of either the singular or the plural form of *campus:* the two seem to have been used indiscriminately, and interchangeably; in any case, as both medieval and modern copyists have found, abbreviation commonly makes a distinction impossible. One example must suffice: the wandering terminology of grants of land at Duntisbourne and Daglingworth (Gloucestershire) in Devine (ed.) (1977), 983-1022.

24. This statement applies likewise to the evidence of some extents. Hilton (1954), 159, writes that 'there are many surveys or terriers of lands in Leicestershire medieval villages which make no reference to this broad division' between two or three fields. In such cases, as with charters which name a profusion of furlongs but no fields, the question must always be asked: is the existence of two or three fields being concealed by the purpose and terminology of the document? Just as one can imagine a grantor finding it more appropriate in a particular case to take the existence of the fields as understood and to be more concerned to enumerate the furlongs within which his land lay, so too one can envisage that, for some purposes, the most useful function of an extent was its measurement of the acreage of demesne in each furlong. Examples of extents which seem to record furlongs according to their proximity one to another, rather than arranging them under 'field' headings, are in Salter (ed.) (1906-8), vol. 2, 2-4 (Shifford, Oxfordshire) and British Library, Cott. Tib. D. vi, f. 72v.-73 (Puddletown, Dorset).

25. But cf. the preceding note. Clearly, differences in the types of formulae used in charters (and in extents) depended upon purpose and local circumstance and upon the availability to those concerned of other documents. An unusual charter of *c.* 1150 (cited by Stenton (ed.) 1920, xxi-ii) spells out at length the grantee's stake in a system involving two fields and it is easy to imagine how the explicit note which it contains could be abbreviated to become one of the

briefer forms more usual in twelfth-century charters.

26. For a possible exception see below, n.40.

27. Royce (ed.) (1892-1903); Hart (ed.) (1863-7); Ross (ed.) (1964); Devine (ed.) (1977). The figures mentioned here in the text are not the result of an exhaustive search, which would no doubt yield additional examples.

28. Postgate (1964), appendix 1; Hallam (1965), 231-6, based on the published collections edited by F.M. Stenton and in the *Registrum Antiquissimum* series. Many of the Lincolnshire charters used here refer to arable holdings equally divided between two 'sides' of a village. That this was the local terminology for two 'fields' is made clear by the following examples: Stenton (ed.) (1920), 232-3; *Registrum Antiquissimum,* vol. 3, 367-8; vol. 4, 69-71; vol. 4, 233-4; vol. 4, 44-6 and 50-1.

29. Stenton (ed.) (1920), 1vii/viii and 140-1; Lennard (1943-5), 137 and Fowler (ed.) (1935), 29-30. For 'inhok' see above, n.19.

30. Gray (1915), 80-1. The totals given here exclude the 'sterile' and 'uncultivated' parts of each field. The source (British Library, Cott. Tib. D. vi, f. 37-37v.) contains details of the new boundaries as they were drawn around the fields, and the marginal note *divisio camporum.*

31. Gurney (1941-6). Another explicit reference to the change from a two-field to a three-field system relates to South Stoke (Oxfordshire) in 1240-1: Gray (1915), 80. An ordinance from Sedgefield (Co. Durham) in 1352 may possibly refer to such a change but is perhaps more likely to be an affirmation of an arrangement already in existence: Ault (1965), 32. In addition, the transformation of one system into the other may possibly be inferred from evidences which seem to indicate a two-field system at one date and a three-field system at a later date: Gray (1915), 76-80; Wake (1922), 406; Chibnall (1965), 221; Beckwith (1967), 108-10; Gurney (1941-6), 251-2 (for a criticism of one of Gray's examples); *V.C.H. Oxfordshire,* vol. 8, 105.

32. For changes to a four-field system after the end of the Middle Ages see, for example, Gray (1915), 132-6 and Havinden (1961), 78-9. An earlier transformation of the same type at Adlestrop (Gloucestershire) in 1498 is discussed by Elrington (1964) while another possible medieval example, at Crowle (Lincolnshire), is mentioned in Thirsk (1964), 22.

33. This case, so often referred to, was first made known by Homans (1941), 56. The source (British Library, Add. MS. 40,010) is a collection of rentals and abstracts from court rolls relating to the estate of Fountains Abbey. The abstract concerning Marton, with a marginal note *divisio campi,* is not dated. A brief inspection of the other materials in the collection does not assist with the dating; I have therefore had to use the dates of the lordship of Richard Scrope who is named in the abstract.

34. Discussed briefly in a later section of this paper.

35. Such arrangements by which the two adjacent fallow fields of two neighbouring townships were brought together during the grazing season (presumably in order to minimise the risk of trespass by animals on the growing crops and to minimise work on making boundaries secure) are interesting as cases of inter-village co-operation in the Middle Ages and may possibly have implications for ideas on the origins of township territories. Examples of medieval agreements relating to the practice are to be found in the following: Stenton (ed.) (1920), 111; Salter (ed.) (1921), 18-20; Gurney (1941-6); Darlington (ed.) (1945), 102-3; Fowler (1925), 83.

36. I know of only two such agreements, those relating to Dry Drayton and Segenhoe, discussed in a later section of this paper.

37. Thirsk (1964), 19. Thirsk does, however, note some evidence from twelfth-century charters of a more regular division of land between two or three

fields but concludes that much of it 'is not sufficient to prove' the widespread existence of mature field systems: ibid, 19-20 and n.27. The contentions of *this* paper are that such evidence is the hall-mark of the Midland system and that it is far more abundant than would be expected if the system had been a novelty in the twelfth century.

38. See also above n.24.

39. This body of material has been surveyed several times in the past, by Nasse (1871, 21-6), Seebohm (1883; 1905 edn, 106-12), Gray (1915, 51-61) and, most recently, by Finberg (1972, 487-97). What is said in this paragraph relies heavily on Finberg's discussion of the material.

40. de G. Birch (1885-93), vol. 3, 617-18; Finberg (1972), 495. Both Finberg and Gray (1915, 60) are more cautious in their evaluation of another Worcestershire lease which refers to 120 acres at Barbourne, 60 to the north and 60 to the south: no fields are mentioned and the cardinal points may well relate to the stream.

41. For example, five hides at Winterbourne (Wiltshire), granted in *c.* 970 or nine hides at Ardington (Berkshire), granted in 961: de G. Birch (1885-93), vol. 3, 395-7, 307-8.

42. West (1969); Addyman and Leigh (1973); Losco-Bradley (1977); Champion (1977). These totally excavated sites, at West Stow (Suffolk), Chalton (Hampshire) and Catholme (Staffordshire), were deserted at an early date; it has yet to be determined whether or not they were entirely typical of early Saxon settlement.

43. Bishop (1935), 23. Bishop's conclusion (p. 24) that 'both the formation and the growth of the open fields were largely preceded by a temporary state of individual clearing and several cultivation' supposes a model for re-occupation of wasted vills altogether different from that suggested by Sheppard (1973) and Harvey (1978) (for which see below, n.70).

44. This composite description is based upon Mayhew (1973), 16-28 and the works there cited.

45. To Finberg the *worthig* of the laws was 'not just a dwelling-place and its yard; the arable, and probably some pasture too, is enclosed within the ring fence': Finberg (1972), 416.

46. The terms are frequent as elements in field names.

47. A reference to 29 *gedale* at Maxey (Northamptonshire) in the tenth century contains no clue as to their acreage: Robertson (1939), 80. Nor can much be made with certainty of the reference to three hides at North Denchworth (Berkshire) which were *undaelede* in 947. This statement may mean that the three hides were not shareland but alternatively may simply imply that they had not been severed from the estate there: Finberg (1972), 492-3; Gelling (1976), 745.

48. This would apply to the type of field system discussed in Harvey (1978).

49. The process by which a deserted settlement might leave its name behind as a field name is mentioned by Beresford in his discussion of sites which lost their habitations in the later Middle Ages: Beresford (1971), 53.

50. One of the fields at Clanfield was called 'Wield', although the name here may derive from the adjacent settlement of Weald: Pocock (1968), 87, 90. For 'Oldfield' and 'Longlands' names see, for example, Pocock (1968), 91-2; Sheppard (1973), 179-80; Ford (1976), 289-91. Such names are very frequent in all pre-enclosure sources relating to field systems.

51. Göransson (1961), 98-102; Homans (1936). In both of these works there are numerous references to medieval holdings where the scattered strips of *x* lie everywhere next to those of *y* or are said to lie *versus solem* or *versus umbram*. Such arrangements could easily come about through the subdivision of a holding strip by strip. In some cases division can be shown to have lain behind the distinctive terminology of the references: for example, Hilton (1949), 30;

Spufford (1965), 22. Much of the evidence commonly cited for 'sun division' in England comes from feet of fines; because a final concord was so often used in the Middle Ages in order to ratify divisions the evidence needs to be used more cautiously than in some discussions of the subject.

52.   At Shirebrook (Derbyshire) a bovate was described as lying *in tribus stadiis camporum*, one next to the park in the South Field, one next to *akkyr hedge in uno fine et villam predictam in campo qui vocatur Tonnefeld*, the third simply *in campo occidentali;* at Kelfield (Lincolnshire) a bovate was described as having nine acres of marsh *in directo terra prenominate* and the whole holding may perhaps be reconstructed as having comprised 'long and narrow blocks of arable stretching . . . from the river over the higher fields beyond': Gray (1915), 461; Stenton (ed.) (1920), 293, xxxiii.

53.   This point is made very clearly in an excellent map, one of the few ot its kind which we have, of types of field systems in the counties bordering Wales: Sylvester (1969), 219-20.

54.   Gray (1915), 107-8. Typical was Gray's treatment of Herefordshire, most of which he included on rather slight evidence, within the zone of Midland field systems. His excellent discussion of post-medieval changes in the county's field systems (pp. 139-53) should for the most part be read as an account not of the decay of the Midland system but of the gradual enclosure of more flexible multi-field arrangements. For a fuller treatment of Herefordshire's medieval field systems, see Roderick (1951).

55.   See, for example, Darlington (ed.) (1945) and Timson (ed.) (1973) *passim;* for a discussion of the evidence on fields contained in the former, Bishop (1946). For the subsequent history of field systems in the countryside bordering the Midland zone see Fox (1975), 199-201.

56.   Somerset Record Office, Ecclesiastical Commissioners, 112968/4; P.R.O. C. 135/260/3.

57.   For example, P.R.O.; C. 133/71/13 (Lydeard Puchardon, Som.); C. 135/32/28 (Sapperton, Gloucs.); C. 135/51/12 (Tewkesbury and Chipping Sodbury, Gloucs.); C. 135/70/7 (Compton Dundon, Som.); C. 145/20/15 (Woolston, Gloucs.); E. 149/9/24 (Yarlington, Som.).

58.   East Devon figures from extents, chiefly those in P.R.O. C. 132, 133, 134 and 135, summarised in Fox (1975), 185. Leicestershire figures also from extents in inquisitions *post mortem:* Raftis (1974), 84.

59.   Because of the survival of the exceptionally interesting documents discussed here these two places both demand far more detailed study. I hope in the future to examine them both, using whatever ancillary documentary and cartographic evidence is available.

60.   Spalding Gentlemen's Society, Wrest Park Cartulary, f. 242-242v.

61.   For the same reason it is unlikely that this series of documents marks a change from a two-field to a three-field system.

62.   'Traces of curving furlongs' have been reported from air photographs of Dry Drayton: *R.C.H.M. West Cambridgeshire*, 1968, 82.

63.   The extent and narrative are f. 7-8 of British Library, Harl. MS. 1885, an early thirteenth-century cartulary of Dunstable Priory. Most of the narrative, but not the extent with which it is intermixed, has been transcribed in Vinogradoff (1892), 457-8. Finberg (1972), 492, made brief use of Vinogradoff's transcription. For another notice see Fowler (ed.) (1926), 252.

64.   The terminology of the narrative does not suggest that it is a digest of charters and other instruments as is sometimes the case with such accounts. Yet a composition from memory seems unlikely. The composition of the cartulary was associated with the compilation of the Dunstable Annals (for which see Cheney, 1969), but the latter contain little material of a domestic nature before 1200 so

that a lost common source is unlikely. The derivation of the narrative must thus remain unsolved for the time being. The dating of the events which it describes is difficult: I may err slightly in the dates which I give here, for which I have used Fowler (ed.) (1926), pedigree 6, *V.C.H. Bedfordshire*, vol. 3 (1912), 104 and some of the sources there cited.

65. Land at Segenhoe called 'Hamstocking', clearly of assart origin, was granted to Dunstable Priory early in the thirteenth century: Fowler (ed.) (1926), 71, 213.

66. Ford has suggested that 'land needed to be intensively cultivated and grazed . . . particularly . . . if any settlement ceased to have access to its traditional pasture': Ford (1976), 292. Jones has suggested that 'the enlargement of former hamlets into villages and the corresponding elaboration of their open fields was one of the major concomitants of the disruption of shires': Jones (1961, a), 200, n.7.

67. Names taken from the list in Gray (1915), 450-509, where the great variety in nomenclature of townships practising the Midland system first set off the line of thought followed in this section.

68. For the critical application to studies of the field landscape of information in boundary clauses of pre-Conquest charters see Gelling (1976), 625-9. For an example of the integration of the study of field systems and estate systems see Phythian-Adams (1978).

69. Cf. Finberg (1972), 491-3, where it is 'very tentatively' suggested that the Scandinavians were a direct influence both on the parcellation of the land into strips and on the 'reorganisation of arable'. The evidence put forward by Finberg is understandably slight; moreover, the overwhelmingly southerly and westerly distribution of the earliest references, in pre-Conquest charters, to acre strips and common arable lends little support to his hypothesis.

70. Bishop (1935); Sheppard (1973), 183-4; Harvey (1978), 21, 24. Between them these works suggest two models for the emergence (or re-emergence?) of regular field systems in those parts of Yorkshire devastated in the campaign of 1069-70. The first (Bishop) is spontaneous recovery followed by assimilation of recolonised land into a regular field system. The second is a planning of field systems as an accompaniment to resettlement, although it should be added that as yet the evidence for planning of villages (Sheppard, 1974 and 1976) has been more thoroughly presented than that for the planning of their fields. For a reconciliation of the two models see Sheppard (1973), 182.

71. That the Midland system and extensive grazing grounds beyond the arable were not altogether mutually exclusive in their distribution is clear from the Northumberland examples given in Butlin (1964), as well as from scattered examples elsewhere.

5 COMMONFIELD ORIGINS – THE REGIONAL DIMENSION

Bruce Campbell

Medieval England possessed a plurality of commonfield systems: yet why this was so, like the related question of commonfield origins, awaits a satisfactory explanation. H.L. Gray was the first to identify and describe different commonfield systems, and he made their existence the keystone of his ethnic explanation of commonfield origins (Gray, 1915; Baker, 1965, a). However, although his regional classification of commonfield systems is still largely accepted, his views on their origin are now discredited. In contrast C.S. and C.S. Orwin (Orwin and Orwin, 1938), in their subsequent hypothesis of commonfield origins, paid little attention to regional variations in field systems and concentrated upon the regular commonfield system of the Midlands. In their view 'wherever you find evidence of open-field farming and at whatever date, it is sufficient to assume that you have got the three-field system at one stage or another' (Orwin, 1938, 127). This preoccupation with the Midland system is echoed in Joan Thirsk's more recent explanation of commonfield origins (Thirsk, 1964 and 1966, 142-7; Titow, 1965; Hilton, 1976, a). In the Thirsk model the Midland system represents the ultimate stage in a long process of evolution, other English field systems reflecting the effects of local and regional peculiarities of environment, settlement history, population density, and agrarian economy, upon the evolutionary process. But is it right to regard the Midland system as the 'norm' from which other field systems deviated?

The Midland system was certainly the most enduring of commonfield systems and for this reason is the system most fully represented in post-medieval sources. But in the fourteenth century, when commonfield farming was most widespread, its pre-eminence was less well marked. A rough estimate on the basis of the 1377 Poll Tax indicates that at most half, and possibly no more than a third, of England's rural population lived in townships whose commonfields were operated according to the Midland system. At least half the remainder lived in townships whose field systems conformed to alternative regimes. Moreover, thirteenth and fourteenth century manorial records leave no doubt that in their medieval heyday the physical development of these alternative field systems often matched, and sometimes even exceeded, that of their Midland counterparts: waste was virtually eliminated, holdings were highly fragmented, and fields were intensely parcellated. (Campbell, 1975). Most significantly, in the medieval period it was the *non*-Midland field systems which coincided with the areas of greatest population density, highest levels of assessed lay wealth, and most advanced and productive agriculture (Brandon, 1972; Campbell, 1975; Darby, 1973; Gray, 1915; Raftis, 1957; Searle, 1974; Saunders, 1930; Smith, 1943).

That being said, the precise nature and exact geographical spread of the various English commonfield systems remain inadequately known. Indeed, no generally accepted set of criteria exists for the identification and definition of different commonfield systems. Some studies have described them in strictly functional terms, but others have utilised a range of functional, morphological, tenurial, and even terminological characteristics (Baker and Butlin, 1973). The resultant lack of a common basis for comparison is a serious defect and it is difficult to resist the suspicion that some of the differences between systems are less real than has been supposed. This suspicion will linger as long as the criteria by which field systems are identified remain ill-defined. Gray's original claim that field systems were unique to different regions is thus still tacitly accepted, and English field systems continue to be classified and described accordingly (Baker, 1965, b).

Consideration of different commonfield systems reveals that they could comprise up to six basic elements: communal ownership of the waste, arable and meadow divided into unenclosed strips, individual holdings made up of a scatter of strips, fallow grazing by the stock of all the cultivators, the disposition of fallow strips controlled by the regulation of cropping, and communal regulation of all these activities. Closer examination of these six main elements allows their refinement, and specification of the following 14 functional attributes:

1. communal ownership of the waste — **THE WASTE**
2. arable and meadow characterised by a combination of closes and unenclosed strips — **FIELD LAYOUT**
3. arable and meadow characterised by a predominance of unenclosed strips
4. holdings made up of an irregular distribution of strips — **HOLDING LAYOUT**
5. holdings made up of a regular distribution of strips

6. full rights of common pasturage on the harvest
   shack
7. limited rights of common pasturage on half-
   year fallows
8. limited rights of common pasturage on full-
   year fallows
9. full rights of common pasturage on half-
   year fallows
10. full rights of common pasturage on full-
    year fallows

} FALLOW
  GRAZING

11. imposition of flexible cropping shifts
12. imposition of a regular crop rotation

} REGULATION
  OF CROPPING

13. seignorial regulation of certain
    collective activities
14. communal regulation of all
    collective activities

} MODE OF
  REGULATION

These 14 items relate to arable field systems, that is to say, field systems where there was a relative shortage of pasture, and provide the criteria upon which a revised functional classification of such field systems can be based. A different set of attributes would need to be specified for pastoral field systems (Dodgshon, 1973; McCourt, 1954-5).[1]

Consideration of the various ways in which these 14 functional elements could be combined (see Table 5.1) allows the identification of five principal types of field system plus several sub-types. These may be characterised as follows:

A. *Non-common subdivided fields* – where common rights are confined to the waste and arable strips are cropped and grazed in severalty. Examples of this type include the subdivided fields of the Lincolnshire Fens and, possibly, those of Kent (Hallam, 1965; Baker, 1965, b).[2]

B. *Irregular commonfield systems with non-regulated cropping* – where fallow strips are subject to rights of common grazing for part, or the whole, of a year, but where cropping takes place in severalty. Here a two-fold distinction can be drawn between:

   (i) those field systems where common grazing rights were confined to the harvest shack (as in eastern Norfolk and parts of eastern Suffolk (Campbell, 1981; Gray, 1915).

   (ii) those field systems where common grazing rights applied both to harvest shack and to land lying fallow at other times of the year (as in south and east Devon (Fox, 1972 and 1975).

C. *Irregular commonfield systems with partially regulated cropping –*
where fallow strips are subject to rights of common grazing, the
disposition of fallow strips is partially controlled by the imposition
of flexible cropping shifts, but where holding layout remains
irregular. These field systems assume three main forms:

(i) where control of a system of flexible cropping shifts is vested in
the seignorial authorities and limited rights of fallow grazing are
instituted, either by restricting fallow grazing to certain types of
livestock (e.g. sheep) or by confining it to certain strips (e.g.
those lying within a particular cropping shift), or by a
combination of the two. The principal example of this type of
commonfield system is the foldcourse system of western Norfolk
and adjacent portions of Suffolk and Cambridgeshire (Allison,
1957; Postgate, 1973).

(ii) where limited rights of common grazing are instituted, together
with associated cropping shifts, and communal control is
established of all aspects of the system (e.g. the field system of
western Cambridgeshire (Postgate, 1964, 1973).

(iii) where full rights of common grazing are instituted, together with
associated cropping shifts, and communal control is established
of the whole system (e.g. the field systems of the Chilterns,
parts of Essex, and the Thames Valley (Roden, 1973).

D. *Irregular commonfield systems with fully regulated cropping –*
where fallow strips are subject to full rights of common grazing, the
disposition of fallow strips is controlled by the imposition of common
rotations, but where, since holdings are made up of a combination of
closes and unenclosed strips, holding layout remains irregular.
Examples of this type of field system include the so-called woodland
systems of the Midlands (Roberts, 1973).

E. *Regular commonfield systems –* where fallow strips are subject to
full rights of common grazing, the disposition of fallow strips is
controlled by the imposition of common rotations, and where
(since unenclosed strips predominate) holding layout is perforce
regular (e.g. the two-, three- and four-field systems of the Midlands
(Gray, 1915).

This functional gradation of field systems poses an obvious problem of
explanation and, indeed, is open to several different interpretations.
Foremost among these is the Thirsk model, according to which common
rights and regulations, like the subdivided fields to which they related

Table 5.1

PRINCIPAL COMPONENTS OF FIELD SYSTEMS

#	Component
1	COMMUNAL OWNERSHIP OF THE WASTE
2	ARABLE & MEADOW CHARACTERISED BY BOTH CLOSES & UNENCLOSED STRIPS
3	ARABLE AND MEADOW CHARACTERISED BY A PREDOMINANCE OF UNENCLOSED STRIPS
4	HOLDINGS MADE UP OF AN IRREGULAR DISTRIBUTION OF STRIPS
5	HOLDINGS MADE UP OF A REGULAR DISTRIBUTION OF STRIPS
6	FULL RIGHTS OF COMMON PASTURAGE ON THE HARVEST SHACK
7	LIMITED RIGHTS OF COMMON PASTURAGE ON HALF-YEAR FALLOWS
8	LIMITED RIGHTS OF COMMON PASTURAGE ON FULL-YEAR FALLOWS
9	FULL RIGHTS OF COMMON PASTURAGE ON HALF-YEAR FALLOWS
10	FULL RIGHTS OF COMMON PASTURAGE ON FULL-YEAR FALLOWS
11	IMPOSITION OF FLEXIBLE CROPPING SHIFTS
12	IMPOSITION OF A REGULAR CROP ROTATION
13	SEIGNORIAL REGULATION OF CERTAIN COLLECTIVE ACTIVITIES
14	COMMUNAL REGULATION OF ALL COLLECTIVE ACTIVITIES

A FUNCTIONAL CLASSIFICATION OF ENGLISH MEDIEVAL FIELD SYSTEMS

	(A)	(B) variant i	(B) variant ii	(C) variant i	(C) variant ii	(C) variant iii	(D)	(E)
		★ ★	★ ★ ★ ★ ★ ★		★ ★	★ ★	★	★
				★ ★				
							★	★
				★ ★ ★	★ ★ ★	★ ★		★
				★ ★			★	★
				★ ★			★	★
			★ ★	★ ★ ★	★			
		★ ★ ★ ★		★ ★	★ ★		★	★
	★ ★ ★	★ ★ ★ ★ ★		★ ★	★ ★		★	★
								★
	★ ★ ★ ★	★ ★ ★ ★	★ ★ ★ ★	★ ★ ★	★			
	★ ★	★ ★ ★ ★		★			★	
	★ ★	★ ★	★ ★	★	★			★
	★ ★	★ ★ ★	★ ★ ★ ★	★ ★	★ ★		★	★

(A) NON-COMMON SUBDIVIDED FIELDS	**(B) Non-regulated cropping** — variant i / variant ii	**(C) Partially Regulated Cropping** — variant i / variant ii / variant iii	**(D) Fully Regulated Cropping**	

IRREGULAR COMMONFIELD SYSTEMS (B, C)

REGULAR COMMONFIELD SYSTEMS (E)

evolved under the impetus of population growth (Thirsk, 1964 and 1966; Baker and Butlin, 1973, 619-56). This led, on the one hand, to an expansion of arable at the expense of pasture, and, on the other, to the proliferation of holdings and strips, thereby simultaneously placing a mounting premium on the temporary forage available on the fallow arable, and creating a need to regularise field and holding layout, so that access by all cultivators to their land and to water was ensured, and meadow and ploughland were protected from stock. To these ends, therefore, holding layout was regularised, communal rotations were introduced to rationalise the disposition of fallow strips, and common grazing rights were established on the latter. As a result better provision was secured for the livestock upon which arable production depended for its traction and manure, without any depletion in the cultivated area. However, this change from non-common subdivided fields to a regular commonfield system was not achieved at a stroke, it proceeded by stages. The first step may have been the establishment of informal agreements between parcenars and groups of cultivators, which in time may have become steadily extended and developed until they became transmuted by custom into formal binding rights. Common grazing of the stubble and aftermath of the harvest may have been the first of these rights to develop for it offered obvious advantages of convenience and was easily instituted, requiring no physical changes in the layout of fields and holdings. Subsequently, the attraction of fallow grazing available at other times of the year coupled, perhaps, with some rationalisation of holding layout, may have led to the adoption of more extended fallow grazing rights. The adoption of such rights may, in turn, have prompted the co-ordination of cropping patterns, so that blocks of contiguous strips remained fallow in the same year and grazing was thereby facilitated. In this way the process of systematisation may have progressed through increasingly developed forms of irregular commonfield system until the ultimate stage, the establishment of a regular commonfield system, was reached. A precondition for attainment of this final stage was a partial or total reallotment and realignment of strips, which could have been undertaken by a village assembly acting in the common interest.[3] However, because systematisation evolved slowly and under different environmental and social conditions, it did not develop everywhere in the same way or to the same extent. In particular, a basic distinction existed between the commonfield systems of populous grain-producing districts, in fertile valleys and plains, and those of less populous districts where there was a greater emphasis upon pastoralism, in upland areas and in the vicinity of extensive forests, marshes and fen. In this way the Thirsk model accommodates regional variations in field systems.

This thesis was first propounded in 1964, since when much evidence has been presented to verify that part of it which deals with the formation of subdivided fields and inter-mixed holdings (Baker, 1964; Spufford, 1965; Sheppard, 1966; Campbell, 1980; Bishop 1935). In contrast, evidence to support the part relating to the genesis of common rules and regulations has been less forthcoming (Campbell, 1981). Indeed, the assumption that increased demand for food led to the progressive rationalisation of the layout of fields and holdings and co-ordination of cropping and grazing practices, can be questioned on two fronts. First there are grounds for questioning the feasibility and advisability of such a course of development. Second, the existence of an alternative and possibly more plausible response to population growth can be demonstrated.

On grounds of practicality, there are several reasons for doubting whether the co-ordination and systematisation of commonfields progressed quite as smoothly, and were quite so directly related to population growth, as the Thirsk model postulates. To begin with, structural innovation (the collective reorganisation of existing fields, holdings and husbandry practices) required a consensus if it was to proceed. Common rights and regulations affected all cultivators and consequently required their unanimous consent before they could be instituted; this applied in particular to the reallotment of land which would have accompanied the regularisation of holding layout and imposition of common rotations. Structural innovation could only take place, therefore, when all the affected parties perceived that it was in their mutual interest. Moreover, to perceive the desirability of structural innovation was one thing: to effect such a change was another, and required a capacity for collective action. Unless communities possessed such a capacity, economic need alone could not have led to structural innovation. But to presuppose that regular commonfield systems were the creation of organised peasant communities begs the essential question whether such communities existed prior to the creation of the commonfield system. They are as likely to have been the effect of the system, as they are to have been the cause.

Even granting the existence of communities capable of making and executing the relevant collective decisions, it remains questionable whether they would have taken the decision to institute a regular commonfield system when peasant numbers were increasing. Such a decision was not to be taken lightly; once executed structural innovation was not readily reversed. This was because it entailed more than the mere physical reorganisation of holdings and fields, complex and controversial though this would have been: at its root lay the institution of a code of inviolable rights and regulations which, once created, could not easily be dissolved without recourse to a higher authority (as in the case of enclosure by Act of Parliament). Furthermore, repeal of these rights and regulations was, like their creation, subject to individual veto. The innate conservatism of peasant cultivators, which was an obstacle to the dissolution of the commonfield system, must surely also have militated against its creation. It is unlikely that individual cultivators would have been willing to agree to an irreversible change of unproven advantage: and they would have been particularly loath to do so when population growth was placing a mounting premium upon proprietorship and leading to a progressive narrowing of the margin of subsistence. Moreover, the larger the number of affected parties, the more remote would have been the prospect of getting them all to agree to structural innovation.

To relate adoption of the regular commonfield system to mounting population pressure presents a further difficulty: the former was innately static, thriving on the maintenance of the *status quo,* whereas the latter was essentially dynamic, and promoted change. Once formed, the regular commonfield system would consequently have been incompatible with continued population growth. If the advantages which the system offered were not to be eroded, therefore, it would have been requisite that the integrity of the new arrangements be preserved by making corresponding changes in the prevailing social and legal code. This accounts for the fact that impartible inheritance and an insistence upon the inalienability of land were invariable concomitants of the regular commonfield system (Homans, 1941; Faith, 1966, Howell, 1976). These, and related customs, served a dual purpose: they not only insulated the system from the divisive effects of population growth, but they also deterred population growth itself. However, the imposition of these customs would probably have met with considerable opposition if it had been attempted whilst peasant numbers were increasing, for, when this was the case, existing proprietory rights and inheritance practices tended to be jealously preserved. Indeed, population growth and customs which favoured the proliferation of holdings tended to be mutually reinforcing and, once established, the momentum which they generated was almost impossible to restrain (Faith, 1966).

More important than these *a priori* objections to the Thirsk model is the fact that it is possible to envisage an alternative, and arguably more plausible, response to the need to raise food production in a situation where holdings were intermixed and reserves of colonisable land had been exhausted. New agricultural methods could have been adopted and known agricultural methods intensified (agricultural innovation and involution). This is an intrinsically natural and practical solution to the need to increase agricultural productivity and several writers have demonstrated that it has been a characteristic agricultural response to population growth in subsistence societies (Chayanov, 1925; Boserup, 1965; de Vries, 1974; Grigg, 1976). Most technological changes could have been introduced within the existing framework of holdings and fields, and would not have required a major reallotment of land or change in regulations. Nor would individual cultivators have needed to obtain the consent of the rest of the commonfield community before they would adopt an innovation; the decision to innovate was an individual, not a collective one. In the absence of common rights and regulations individuals would have been under no obligation to make their husbandry practices conform with those of other cultivators, and the intensity, and the techniques of agricultural production could therefore have been adapted to individual circumstance. Experimentation was possible, in most cases changes were reversible, and technological innovation could thereby progress by relatively safe and easy stages. Since change could take place gradually, across a broad front, the risk attached to adopting any one innovation was minimised: the maintenance of an artificial equilibrium was not at stake.

Of course, technological innovation would not have been possible unless the necessary technology was available, and this has been the subject of considerable debate (White, 1962; Fussell, 1968; Titow, 1969). Some writers have represented medieval agriculture as technologically inert and incapable of increasing arable production without jeopardising soil fertility and precipitating falling yields (Titow, 1972). Yet there is a growing body of evidence to show that the necessary technology, by which agricultural production could be intensified, did exist, and it seems likely that this was employed with considerable success in several localities in eastern and south-eastern England by the close of the thirteenth century (Richard, 1892; Slicher van Bath, 1960). The technological means by which sustained increases in output per unit area were attained in these localities were manifold, but at their core lay the substitution of fodder crops for bare fallows and natural pasture. In the thirteenth century the principal fodder crops were peas and beans, although vetches were also known: oats, too, were grown as a fodder crop. To derive maximum advantage from this change to feed produced from natural fodder, horses were substituted for oxen in ploughing on account of their greater speed and capacity to work longer hours. Horse ploughing and harrowing also allowed more thorough preparation of the seed-bed, which was further improved by ploughing in farmyard manure from stall-fed livestock. Although considerably more labour-intensive than the pasturing of livestock upon the fallow fields, the manual spreading of manure was far more effective, as it reduced the twin losses from leaching and oxidisation and ensured much more even coverage. The importance attached to the maintenance of soil fertility is further reflected in the assiduous use made of all available supplies of fertiliser, including the folding of sheep upon isolated fallow strips, and the spreading of marl and urban refuse where available. The cultivation of leguminous fodder crops also enhanced soil fertility by raising its nitrogen content, whilst a heavy emphasis upon spring-sown crops allowed regular fallowing on a half-year basis. Flexible rotations, thick sowings, and careful weeding and harvesting also contributed to the productivity of this remarkably intensive system of husbandry. Its rewards were the near elimination of bare fallows, an expansion of the cultivated area to the maximum extent possible, and the attainment of high and sustained yields per acre (although often at the expense of only moderate yields per seed).[4] Where these technological innovations were made, as in eastern Norfolk, irregular commonfield systems with a minimum of common rights and regulations successfully supported exceptionally high population densities (Campbell, forthcoming).

By the thirteenth century, therefore, technological innovation and involution had become an extremely viable and effective alternative to structural innovation. The association of very progressive agricultural methods with highly irregular field systems also shows that chaos and inefficiency were not inevitable corollaries of a failure to rationalise the layout of holdings and fields and co-ordinate their cultivation and grazing. In fact, there were a number of localities in the thirteenth century which possessed all the preconditions which should supposedly have led to the adoption of a regular commonfield system; yet such a system was not adopted because the alternative response of technological innovation had been followed (Campbell, 1980; Baker, 1965, b; Gray, 1915, 302-3, 331-2). Indeed, to a considerable extent these two methods of raising food production were mutually exclusive. Innovations such as the substitution of fodder crops for bare fallows, flexible rotations, and the stall-feeding of livestock, would have been incompatible with a fully regularised commonfield system.

Although technological innovation was arguably the more viable response to population growth, structural innovation was of course adopted in some measure in many localities. This, however, is less likely to have been derived from population growth than from population stagnation and decline. Research into population trends prior to the demographic transition of the nineteenth century has now established that numerous short-term fluctuations in population were superimposed upon a long-term sequence of demographic 'cycles' (Helleiner, 1967; Wrigley, 1969). After a long initial phase of sustained population growth, each of these 'cycles' comprised a period of prolonged demographic contraction, heralded by a brief intermediate phase of mounting demographic crisis. Thus, sustained population growth during the twelfth and thirteenth centuries culminated in a period of acute demographic crisis during the first half of the fourteenth century: approximately a century and a half of population decline and stagnation then ensued (Hatcher, 1977; Miller and Hatcher, 1978). The next demographic 'cycle' commenced in the first half of the sixteenth century. Rapid population growth during the sixteenth and first part of the seventeenth centuries was terminated by renewed demographic crisis in the second quarter of the latter century; this ushered in a further century of demographic malaise, comprising stagnation in some localities, contraction in others (Smith, 1978). Finally, a last wave of sustained population growth during the second half of the eighteenth century culminated, not in demographic crisis, but the demographic transition, and the establishment of a new demographic pattern (Flinn, 1970). This 'cyclic' pattern was the product of important temporal shifts in mortality and fertility rates brought about by the complex interaction of biological and economic forces (Wrigley, 1966, 1968, 1969; Chambers, 1972). Given the demonstrable existence of this pattern from the mid-twelfth to the late eighteenth centuries, therefore, it is *a priori* likely that earlier periods also experienced successive waves of population growth separated by long intervals of demographic recession. As yet little is known of the chronology of population trends prior to the twelfth century, but the evidence of the Irish Annals, the Anglo-Saxon Chronicle, and other contemporary accounts does point to a number of periods of heightened mortality, and consequently, perhaps, of demographic decline, notably in the late seventh century, the closing decades of the eighth century, and again in the late eleventh century (Creighton, 1894; Bonser, 1963; Howe, 1972).

Such variations in the size of populations have obvious implications for the development of field systems and other agrarian institutions. But whereas the importance of periods of rising population has long been appreciated, especially with regard to the fragmentation of holdings and parcellation of fields, the influence of periods of stagnant and declining population has yet to receive serious consideration (except Fox, 1972, 1975). Such periods may, however, have made an equally important, if different, contribution to the development of field systems. It is arguable that they were peculiarly conducive to the rationalisation and systematisation of existing fields, holdings and husbandry practices. A principal reason for this was that as populations fell so it would have become advantageous to pool scarce labour resources in order to use them more efficiently. Establishment of a regular commonfield system enabled this; substantial labour savings derived from conducting activities in common which had previously been undertaken in severalty. The common grazing of fallow strips is a case in point. By combining common rotations with common grazing rights it was possible to pasture entire furlongs and fields with a minimum of supervision. The alternative, under a system of farming in severalty, was to fold or tether livestock on individual, scattered fallow strips, a practice which required the fencing of adjacent strips against the depredations of the livestock, should they have got loose. Further economies in labour were afforded by the greater convenience of a regular layout of holdings and fields.

Circumstances of falling rather than rising population would also have presented fewer practical problems to the rationalisation and systematisation of commonfields. Where the population was falling there would have been no pressure to intensify methods of cultivation and hence fewer objections to adopting a permanent rotational scheme which provided for the fallowing of blocks of strips on a regular basis. On the contrary, there would have been much to recommend a device which helped to maintain existing levels of productivity. Moreover, where the demand for land was slack it would have been fairly easy to effect changes in the layout of fields and holdings by the exchange and consolidation of strips, and, since there was little pressure upon land and all cultivators stood to derive advantages from this exercise, it can be presumed that few would have objected to it. In fact, the more the number of cultivators declined the more likely would have been the prospect of securing the consensus necessary for structural innovation. Circumstances would also have favoured the social changes consequent upon the institution of a regular commonfield system. Where younger sons were few, and there was no shortage of holdings, there would have been little hardship in making the changeover from partible to impartible inheritance and an insistence upon the inalienability of land (Campbell, forthcoming).

These general considerations still leave unexplained the marked regional variations in the character of structural innovation. Obviously, regional variations in population may have some bearing on this matter. Areas of relatively high or relatively low population density, for instance, may have been less conducive to structural innovation than areas of moderate population density. This is because if the population density was too high the obstacles to structural innovation may have been insurmountable, its advantages may have been of little relevance, and technological innovation may have been more appropriate. Conversely, if the population density was too slight there may have been little need for rationalisation and systematisation: rather than a trend towards more developed forms of commonfield system there may have been reversion to consolidated holdings and enclosed fields held in severalty, accompanied by the adoption of a more extensive form of agriculture. This would account for the fact that the regular commonfield system was absent from the most and the least populous areas. However, although the distribution of population undoubtedly exerted a general influence upon the distribution of regular and irregular commonfield systems, it cannot account for the detailed distribution of the different functional types of field system. Other conditioning factors almost certainly existed.

In this connection, and granted the doubts which have been expressed about the capacity of peasant communities to undertake structural change, considerable interest attaches to the institutional factor of lordship. Lordship was subject to important geographical variations, and three main ways may be envisaged in which variations in its authority, structure and continuity could have determined regional differences in field systems. First, the existence of peasant communities with a capacity for acting collectively may have been a function of strong and undivided lordship. Second, although peasant communities may have been the prime movers, lords may have been the instruments of structural innovation: in other words, once a peasant community had taken the relevant decision it may have referred the matter to its lord, on account of his superior authority, for implementation. Third, the creation of a regular layout of holdings and fields and the institution of common rights and regulations may have been a direct, seignorial imposition. In each of these three cases, it should be noted, strong and undivided lordship would have been most favourable to the functional development of the commonfield system.

Little is as yet known about the development of peasant communities as corporate entities, but it is plain that a capacity for collective action could have been of primary significance to the creation of centrally organised, peasant farming systems. This capacity could have originated in a number of ways, not least as a response to seignorial authority. Thus, peasants may have derived an initial sense of shared identity from the protection afforded by their lord, an identity which would have been reinforced by the legal bonds which tied tenant to lord. Progressive subordination of peasants by their lords would subsequently have prepared the way for collective action by the removal of social differences between peasants and the creation of a homogeneous class of customary tenants. The potential for collective action may then have been lent cohesion by, and found expression in, peasant resistance to seignorial exploitation and oppression. In this way peasant communities may have crystallised as corporate entities. Given that strong and undivided lordship was fundamental to this process, it follows that there may have been an important indirect relationship between spatial variations in the nature of lordship, and spatial variations in the organisation of peasant farming systems.

The impact of lordship upon the development of field systems may, however, have been more direct than this. It is possible that commonfield systems were the product of co-operation between lords and their tenants. The complex and controversial tasks of rationalising the layout of fields and holdings and instituting common rights and regulations may have been referred to lords by their tenants. After all, the lord would have been better placed to make an impartial *remembrement*. He had the authority necessary to execute this task and he probably had readier access to the requisite skills of valuation and surveying, and his court provided an established framework for the subsequent enforcement and regulation of the system. For lords to have been capable of acting in this way, however, it was necessary that their jurisdiction extend to the entire vill and its inhabitants. Again, therefore, circumstances would have been most favourable to the functional development of field systems in vills of strong and undivided lordship.

It is entirely, plausible, however, that lords intervened in the agricultural practices of their tenants for less altruistic reasons. Exploitation may have been their motive. Under the feudal mode of production peasants were personally unfree and legally subordinate to their lords, who were empowered to appropriate their surplus labour product (i.e. the labour and goods which were surplus to the satisfaction of the immediate subsistence requirements of the peasant and his family (Dobb, 1946; Duby, 1974; Hilton, 1976a). This surplus labour product was appropriated either in the form of direct labour services on the lord's demesne or in various forms of rent. Lords thus had a vested interest in the agricultural organisation of their tenants. Where labour was relatively scarce, therefore, they stood to gain from a more efficient deployment of peasant labour. As already observed, imposition of the regular commonfield system had precisely this effect; by pooling labour resources and arranging for certain activities to be carried out in common it facilitated the release of labour to work on the lord's demesne. Division of the demesne into strips located in the commonfields furthered this arrangement. The whole scheme may therefore have originated with the lord, tenants having no alternative but to comply. On the other hand, where labour was relatively abundant it would have been unnecessary to go to these lengths to secure the necessary labour to work the demesne, and recourse may have been made to alternative methods of appropriating the peasants' surplus labour product. The juridical authority vested in the lords allowed them to act in this autocratic manner, whilst the creation of standardised customary holdings furnished them with a suitable opportunity. Since the standardisation of holding size and regularisation of holding layout would both have required a substantial reallotment of land, both could have been instituted together. Indeed, the fact that virgates and bovates were agrarian as well as tenurial units implies that this was so. If the systematisation of commonfields was a function of the development of the feudal system, it again follows that it would have progressed furthest where there was a tradition of strong and undivided lordship and where there was a high propotion of customary tenants.

It thus transpires that there was a variety of ways in which differences in the authority, structure and continuity of lordship may have exerted a determining influence upon the development of field systems (Goränsson, 1958; Smith, 1967; Sheppard, 1976). In all cases this influence appears to have been most formative where a lord's territorial jurisdiction encompassed an entire vill, and where a substantial majority of the inhabitants of that vill were subordinate to the authority of the lord. Circumstances would have been less propitious for the rationalisation and systematisation of commonfields where only one, or neither, of these conditions prevailed. In such vills the capacity of the lord to influence the entire community, or to impose his will on the entire vill, would have been correspondingly reduced. Other things being equal, a distinction should consequently have existed between the field systems of areas where lordship was strong and those of areas where it was weak.

The evidence by which this supposition may be tested is limited because knowledge of the distribution of lordship and distribution of field systems is defective. Nevertheless, the evidence of Domesday Book, the Hundred Rolls, and the *Inquisitiones Post Mortem* does suggest that within lowland England lordship was consistently stronger in some areas than others (Kosminsky, 1956; Darby, 1977; Miller and Hatcher, 1978). In particular, the central and southern counties (the heartland of the old Anglo-Saxon state) appear to have been characterised by a higher incidence of vills of undivided lordship, and a lower incidence of freemen, than the eastern and south-eastern counties (where the continuity of lordship was disrupted by the Scandinavian incursions of the ninth and tenth centuries). Significantly, it was in the central and southern counties that regular commonfield systems were most fully developed, and in the eastern and south-eastern counties that various forms of irregular commonfield system prevailed. This general coincidence between areas of strong lordship and regular commonfield systems, and areas of weak lordship and irregular commonfield systems, also shows up at a more local level. Thus, in the heart of the Midlands, a fundamental contrast existed between the two adjacent Warwickshire hundreds of Kineton and Stoneleigh (Harley, 1958; Roberts, 1973). In the late thirteenth century the former was an area of relatively old settlement, strong lordship, and regular commonfield systems: its population evinced remarkably little increase during the twelfth and thirteenth centuries and contained very few freemen. The latter, on the other hand, was an area of more recent settlement; lordship was less firmly established, freemen were better represented, and many townships recorded a substantial increase in population during the twelfth and thirteenth centuries. In this hundred an irregular commonfield system with fully regulated cropping prevailed. A similar dichotomy may be seen in Norfolk. In the west of the county an irregular commonfield system with partially regulated cropping was the norm, whereas in the easternmost hundreds field systems were even less systematised and an irregular commonfield system with non-regulated cropping existed. This reflects the fact that the seignorial nexus was even weaker in the east than it was in the west; lordship was characterised by a higher degree of fragmentation and the freeholding element in the population was more substantial (Douglas, 1927; Darby, 1952; Allison, 1957, 1958; Rainbird Clarke, 1960; Campbell, 1981).

These examples suggest that if the functional gradation of field systems is explicable in terms of a single factor, that factor may be the structure of lordship: the greater the authority and continuity of lordship, the more fully systematised the commonfield system.

Nevertheless, it has not been the purpose of this paper to present a new monocausal explanation of commonfield origins. Rather, the intention has been to clarify definitions, remove preconceptions and pose questions. In this context the plurality of commonfield systems has been of central concern, for it is upon this issue that current explanations of commonfield origins founder. Only if the commonfield system is studied in its entirety and in its full socio-economic context will it be possible to account for the temporal processes which gave rise to this spatial complexity. As has been argued, such diverse variables as technological innovation in agriculture, the course of population change, and the institutional framework of society may have been crucial to the functional development of field systems. Yet current knowledge of each of these variables, particularly during the pre-Conquest period, is sadly deficient. So too is our knowledge of regional variations in field systems. Greater knowledge of the variant forms of commonfield system which existed, of their precise formation, their distribution, and their development, is fundamental to the explanation of commonfield origins.

Notes

1.   Pastoral field systems were characterised by a shortage of arable land and a relative abundance of pasture and were principally distinguished by periodic cultivation of the common waste.

2.   Although as yet unproven, it remains possible that the subdivided fields of north and east Kent were subject to rights of common grazing on the aftermath of the harvest.

3.   The Thirsk model also postulates that the piecemeal buying, selling, exchanging and leasing of land may have been alternative means by which holding layout became regularised.

4.   In the first half of the fourteenth century the mean yield of wheat and barley on the most productive demesnes might be as high as 20 bushels per acre, rising to over 30 bushels per acre in a very good year. The equivalent mean yield ratios of these two crops were 6 fold and 3½ fold respectively.

6 THE INTERPRETATION OF SUBDIVIDED FIELDS: A STUDY IN PRIVATE OR COMMUNAL INTERESTS?

Robert Dodgshon

The subdivision and intermixture of property — or what Alan Baker has termed for us subdivided fields (Baker, 1969, 139) — forms the minimum or essential characteristic of any open-field system. Without it, other attributes, such as the communal regulation of cropping or rights of common grazing over arable after harvest, lose their *raison d'être*. But while the debate over how subdivided fields originated has expanded impressively, it cannot be said to have advanced forward, since although we now have a clearer definition of what they comprised and more evidence concerning the credibility of particular interpretations, we are no nearer a consensus over which interpretation best explains them. In short, the problem has gathered viewpoints rather like a boat gathers barnacles. The growing encrustations of opinion have slowly brought it to a point at which — while still eminently seaworthy — it tends to appear slow and cumbersome in its movement. So much so, that to review current opinion on subdivided fields is, in effect, to review *all* opinion as far back as the pioneer contributions by scholars like F. Seebohm and P. Vinogradoff.

Yet paradoxically, this dilemma may reflect the intrinsic nature of the problem rather than our uncertain and conflicting efforts to understand it. As I argued in a paper published in 1975 (Dodgshon, 1975, a, 3-29), and as I propose to reiterate in this paper, it may well be that far from having a situation in which scholars have been persistently talking at cross purposes with each other, more than one interpretation may indeed be valid. Once this is accepted, the task facing any review of the problem alters. So long as we seek only to acknowledge a single interpretation as more valid than the rest, to make a choice between possibilities, then the discussion automatically concludes itself once it has established the superior claims of that interpretation. However, to offer a mix of interpretations — not from uncertainty but from conviction that this is necessary — creates new, additional problems. We need to explore the ways in which the interpretations that comprise this mix were compatible. I propose to do this in two ways. First, I want to show how their co-existence can be rationalised by linking them to specific forms of subdivided field — forms that have come more to the fore in recent years — or to different stages in the process by which such forms were produced. Secondly, I want to show how any remaining question marks over their compatibility can be removed altogether by structuring them into a single, compound interpretation, one that adapts itself to historical and geographical changes in circumstance. In many ways, I feel this latter issue is the most challenging problem arising out of how subdivided fields developed. To acknowledge the relevance of more than one interpretation tempts dismissal as a facile stratagem for circumventing the current conflict of viewpoint, one which, on any inspection, is hardly the easiest of arguments to justify. It makes much more sense to see such an idiosyncratic form of field layout as proceeding from one rather than a group of causal factors. What I wish to propose is that this difficulty recedes when we consider whether the interpretations that comprise our chosen mix were linked functionally together, with each drawing on a common idea but tailored to meet changing circumstances.

Interpretations of Subdivided Fields

Altogether, there are five different interpretations of how subdivided fields developed to be found in the literature. All have been part of the debate for many years in one form or another, with scholars re-discovering or re-stating their support for them at periodic intervals. In the brief review that follows, I hope to draw out this longevity of life that has characterised them because it highlights the point I made earlier that, for all the debate expended on this subject, there has been surprisingly little movement of ideas.

The first interpretation of any substance was that advanced by F. Seebohm. It is to Seebohm that we owe the earliest formal statement of the view that subdivided fields developed out of the practice of co-aration or joint ploughing using a heavy mouldboard plough. Under such an arrangement, landholders subscribed to a joint or common plough-team, providing one or more oxen for the team or parts for the plough itself. This subscriptive system of ploughing was seen as creating the need for a division of ploughed land between those involved. It appeared logical to Seebohm that the method of allocation adopted by such communities should be based on what a plough-team could manage in a single day — a unit of work that formed the strip or selion — and that each of the landholders who contributed to the team should then receive a strip or day's ploughing in sequence. Thus, as the plough-team worked its way across the arable land of the township, landholders built up a portfolio of scattered strips as their share (Seebohm, 1883, 117-25). Subsequent scholars, such as the Orwins, have tightened some of the assumptions implicit in this argument. The Orwins, for instance, drew attention to the fact that the technical restraints of a *fixed* mouldboard plough made it imperative to adopt the strip as the unitary form in which land was ploughed (Orwin and Orwin, 1938). Once this was appreciated, then the use of such strips as the unit of allocation seemed both sensible and desirable. It is this element of technical determinism that has ensured the strong survival of this interpretation as an acceptable possibility, with historians generally feeling that its reasoning dealt with tangibles rather than the somewhat nebulous qualities of its rival interpretations (Postan, 1972, 129). Thanks largely to this acceptance by many influential historians, it has become the view for popular consumption, a status decorated, if not endorsed, by the way Monty Python wrote it into a recent song of theirs called 'The Open Field Farming System'.[1] Those of you who know the song in question will understand me when I say that as someone of part Welsh stock, who has spent much of his research time working on Scotland, I feel great empathy with a certain Professor Angus Jones of the Open University, but I care not to sing along with him when he expounds the seminal importance of joint ploughing with an eight-ox team to the formation of the open field system.

My doubt springs from one basic weakness of the interpretation. Put simply, it has a persuasive logic to it but it lacks explicit support from historical evidence. Seebohm himself cited extracts from the Welsh law codes which formalise the order or sequence by which landholders who contributed to a plough-team were to receive their strips or *erwau* (Seebohm, 1883, 120-2). To the unwary, this may seem unequivocal proof that at a very early date joint ploughing produced subdivided fields. However, care must be taken to distinguish cause from method. Seebohm's extracts only conclusively prove that the different contributions to each joint plough-team were used to fix the order of land allocation. It does not demonstrate conclusively that this was also the *cause* of division. To those unfamiliar with the mechanics of land division, this may seem a fine distinction to draw, but is one emphatically made by evidence for other areas and one that I will discuss in more detail later. In fact, one can find areas where joint ploughing was widespread, yet where it was neither cause nor method as regards the laying out of subdivided fields (Thomson, 1919, 318;

Dodgshon, 1980).[2] Other scholars who have lent their support
to this interpretation have failed to overcome this crucial deficiency.
Until this can be done, we are forced to pass an open verdict on it as a
cause of subdivision, if not on its use as a method of land allocation or
its use as a form of co-operative husbandry.

Soon after Seebohm had published his ideas, P. Vinogradoff put
forward an alternative argument. It was based on the idea that the
bovates and virgates out of which the average English township was
constructed and by which each landholder's portion was calculated
denoted aliquot shares in the township. More important, each share was
construed as equal in both extent and value, measure for measure. If
land was of variable quality (as it invariably was in most townships),
Vinogradoff saw no way in which such shares could be divided equally
using these criteria without giving each landholder a share in all the
different sectors or furlongs. The end result was subdivision
(Vinogradoff, 1892, 235-6; 1905). Although he developed his ideas
through a number of publications, Vinogradoff was unable to tie cause
to effect in a convincing way. Even with the formidable support of
F.W. Maitland, no unambiguous evidence was cited showing that
landholders were trying to equalise shares when laying out subdivided
fields (Maitland, 1960, 404). The argument seemed an exercise in
inference, no doubt guided by Vinogradoff's knowledge of Russia
where the evidence for shareholding was more clearcut.

In fact, it was only recently that evidence was published demonstrating
how tenure, in the form of shareholding, could be relevant. The work I
refer to is my own on Scotland. Scottish rentals and tacks for the
period before 1750 abound with instances of fermtouns that were held
by more than one tenant, and in which each tenant held an explicitly
defined share, such as a quarter, a half or a third. Intermixed with them
were other touns in which each tenant held a share expressed in terms
of land units like merklands or husbandlands. The fact that some tacks
or rentals treat both forms of description as alternative styles for the
same tenure affirms Vinogradoff's assertion that land units like oxgate
and bovates (both of which occur in Scotland on a small scale) could
have the meaning of shares, rather than holdings fixed and defined on
the ground. But of greater interest is the documentary evidence which
links the holding of land in shares directly to the formation of
subdivided fields or, as they were called in Scotland, runrig. Early law
books from the seventeenth century onwards as well as court
proceedings exist which place the link between the two beyond doubt,
at least in a Scottish context. Only when tenants or heritors had divided

their shares into actual holdings on the ground could they begin farming. Such sources also make it clear that where we are dealing with a case of shareholding, then we need to establish more than simply the cause and method of division. We also need to establish the objectives of the division, or the principles guiding the interpretation of shares. One or two land charters offered landholders the choice of having their share in the form of a consolidated holding or in the form of runrig. Law books confirm that this was an option *always* open to shareholders or portioners. The reason that landholders opted for runrig layouts, therefore, cannot be taken for granted. It must be explained *through* the interpretation of shares as shares. Fortunately, there is evidence for this aspect of the problem. Many tacks and land charters describe shares as being 'just and equal' portions of a toun, a phrase which by itself implies that their layout was still a matter for debate. But what matters here is just how equal? Eighteenth-century court records for south-east Scotland contain division proceedings for runrig touns in which the interpretation of shares is fully discussed. They reveal the critical factor to have been exactly what Vinogradoff had reasoned, or an equality of both extent and value between shares (Dodgshon, 1975c). Indeed, when the Perthshire runrig toun of Inchyra was discussed in the House of Lords back in 1876, at least one peer used comparable evidence to sustain an identical conclusion.[3]

Closely linked with shareholding as a cause of subdivided fields is the idea that the practice of partible inheritance had a formative influence. H.L. Gray was the first to introduce this notion in his book *English Field Systems* (1915). It was, he asserted, the cause of subdivided fields in Celtic areas and in Kent. So potent a force was it in the former, that he talks of whole townships being reduced within a few generations to complex subdivided systems. Yet he considered the effect of partible inheritance as only an accidental cause of subdivision. Townships became subdivided by chance not by design or intent. What he meant by this was that it arose from a conjunction of circumstances at an unforeseen point, not because their underlying constitution as farming townships determined that it should be so (Gray, 1915, 199-202). But in so far as Gray was suggesting that an equal partition of land between co-heirs could produce subdivided fields, he was really offering a specific instance of Vinogradoff's shareholding hypothesis, one of a number of reasons why the proprietary or tenurial structure of a township could be reduced to a shared basis. However, unlike Vinogradoff, he glosses over why such shares produced subdivision. We are left to presume that he had in mind the stress on absolute equality with which Vinogradoff had infused his argument on shareholding.

Since Gray wrote, the case for partible inheritance as a specific cause of subdivided fields has been strengthened, but not in the way he might have expected. Without denying that there are studies of Celtic areas which lend firm support to his ideas, there are others from which support is not forthcoming (Jones, 1973, 449-51; Clouston, 1918; Mackay, 1897-8).[4] Perhaps more important, there are now scholars like Joan Thirsk who are prepared to see partible inheritance as having a determinant effect on the formation of subdivided fields even in Anglo-Saxon areas like the Midlands (Thirsk, 1964, 11-14). However, stepping back from this question of how extensive its influences were, what matters most for the argument in hand is that detailed case studies — such as that by Baker in Kent (Baker, 1964, 1-22) — have now floated this interpretation in a securely watertight form.

An attractive interpretation of how subdivided fields were formed, but one which seemed destined until recently to be dealt with implicitly rather than explicitly, was that which invoked the piecemeal colonisation of land by communities. The township is conceived as a chequerboard on which each square is colonised step by step, with the various holdings that made up the township acquiring a personal foothold at each new step. In its most orderly form, we can imagine the original nucleus of holdings as forming a sort of DNA code that imprinted itself or its character on each new growth. But orderly or disorderly, on the initiative of the community or individuals, such a piecemeal growth had the means to fragment the rights of each landholder or holding across the entire resource network of the township. Seen in this way, subdivided fields take on the character of an institution that evolved, but slowly rather than one which was devised overnight in a single act of planning. Most discussions of subdivided fields have, in fact, reserved some place in their argument for the problem of colonisation or growth. To a large extent, though, it has tended to be used as everyone's Cinderella. It performs much of the drudgery in the formation of subdivided fields, but is never for one moment thought of as having pride of place as cause. Seebohm, for example, thought it important to his argument that as new land was taken into cultivation, new plough-teams could easily be added since the allocation of shares was bound up with the process of ploughing (Seebohm, 1883, 113-14). The Orwins gave colonisation slightly more importance, but they too held back from seeing it as a possible *primum movens* (Orwin and Orwin, 1938, 42). It contributed greatly to the complexity of subdivided fields, but was not a primary cause of them. But more recently, the discussions by

Alan Baker and Robin Butlin (Baker and Butlin, 1973, 637-9) and by R.C. Hoffman (Hoffman, 1975, 54-5) have reassessed its position reinstating it as a cause that can no longer be taken for granted. For proof, we need only witness one phase of colonisation during which land was shared between landholders to realise that a sequence of such phases could fragment and interject landholding. Studies like H.E. Hallam's work on the Lincolnshire Fens is rich in examples (Hallam, 1965, 29; Dodgshon, 1975c). Ironically, it is the work of T.A.M. Bishop on the connection between assarting and subdivided fields that has tended to mislead us over the true potential of piecemeal colonisation. He stressed how colonisation led initially to parcels held in severalty. Only later did these parcels become subdivided into strips (Bishop, 1935, 13-29). Other, separate factors, therefore, seemed ultimately responsible for the critical shift into subdivision. Without quibbling that what Bishop describes is a specimen of the processes involved, what he says needs to be seen in a wider context. Basically, what matters is that holdings should become interjected *at any scale*. Once they had acquired even disjoined *blocks* of land in any new sector of colonisation, then arguably the critical shift has already been made and the conversion to smaller units of intermixture is a matter of degree, or at least a change within a mould already cast. This is a theme on which the discussions by Baker and Butlin and by Hoffman focus. Piecemeal colonisation may have been a primary process, but it did not operate in an exclusive or pure fashion. As it proceeded, the pattern it created was subjected to all sorts of elaborative influences, such as the partition of family holdings or sales, influences which served to re-work blocks of landholding into finer-scaled units of intermixture (Baker, 1964-5, 1-22 and 1973, 393-401).[5]

In one of the more searching analyses of the problem, D. McCloskey has presented the case for subdivided fields being an instance of risk aversion, an insurance against crop failure or destruction. It was 'a Good Thing', whereas 'a consolidated holding was a hazardous holding' (McCloskey, 1973, 154). Ideas that subdivided fields conferred a share of the good and bad, wet and dry, on each landholder have long been part of the debate over their origin, but it was McCloskey who first elevated them to the level of a serious proposition. Of course, there is far more to his argument than mere suggestion. Indeed, few other studies can compare in their depth of analysis. But there is doubt over what exactly he proves. He most certainly demonstrates that subdivided fields could function as an insurance against risk, but it is another matter to assume that this was also cause and not just effect, or that it

was a genuine case of behaviour against risk. We can only be assured of
the latter if it is read from the process and not just the pattern. The
data available does not permit him to do this. Moreover, he inclines to
gloss over the deficiency. This is apparent from the way he criticises the
alternative interpretation based on partible inheritance. He suggests that
while peasants were 'cautious', they were not necessarily egalitarian
when it came to dividing their patrimony between heirs. Yet what he
refers to is the division into shares, not the way those shares (however
disproportionate in size they may have been) were then laid out as
holdings (Hoffman, 1975, 162). The two were not the same. Significantly,
this is precisely the area of the problem on which his own argument
fails to convince. There is abundant and satisfying evidence of
McCloskey thinking his way from pattern back to process, but very
little for peasants thinking along the same lines but in the reverse
direction. This comparison between his own interpretation and that
based on partible inheritance has a further purpose to it. Had he
explored the principles which shaped the laying out of shares between
co-heirs, he might have found it difficult to disentangle them from
those to which he himself appeals. The point I have in mind is that
what a lawyer might express as an equality of extent and value, a
farmer might see as an equality in the spread of risk and profit. Here,
somewhat re-focused, may be the real contribution of his discussion.

The Compatibility of Interpretations

I have tried to underline in a general way the plausibility of not just one
but a group or mix of interpretations. For some scholars this sort of
open verdict has proved unacceptable. Their response has been to
override the case of supposedly weaker views in order to assert the
claims of one alone. However, some have not done so without their
misgivings. Some, for instance, have found a place in their arguments
for secondary or contributory causes. Thus, the Orwins saw co-operative
ploughing as the primary cause, but piecemeal colonisation as
contributory (Orwin and Orwin, 1938, 39, 42). Hoffman saw piecemeal
colonisation as the primary cause, but partible inheritance as
contributory (Hoffman, 1975, 54-5). Similarly, those who have
advanced partible inheritance as the primary cause have tended
implicitly or explicitly to enjoin the principle of shareholding to their
argument. A few, and only a few, have responded to the situation by
upholding the claims of two equally sound interpretations. Gray, for

example, gives the impression that he thought both partible inheritance and joint ploughing were viable (Gray, 1915, 199). More recently, Baker and Butlin have offered their unequivocal support for both partible inheritance and piecemeal colonisation, since both appeared to work (Baker and Butlin, 1973, 635-41).

Contrary to what many past scholars have lightly assumed, I believe there is a virtue in favouring more than one interpretation as Baker and Butlin have done, and Gray before them. The circumstances surrounding subdivided fields demand an interpretation of breadth and flexibility. A mix of interpretations has this greater explanatory power. But apart from our conspicuous failure to agree on one interpretation alone, there is another reason why we should settle on a mixed solution as a realistic possibility rather than as a makeshift position to cover our inconclusiveness. What I have in mind is the sheer variety of layout that characterises subdivided fields. The full definition of this variety has been one of the achievements of the many admirable local and regional studies of field systems. The patterns discerned range from those possessed of an impressive regularity in the size and layout of strips to those dominated by irregularly-disposed parcels of widely differing sizes and shapes. One can also distinguish between systems whose every sector was subjected to subdivision and those in which subdivision was more confined. Amongst the latter, we can distinguish between townships in which there was no particular pattern to the areas subdivided and those in which they formed the nucleus of the township, a nucleus encircled by a fringe of several parcels or enclosures. Broadly speaking, we can reduce this variety to a simple classification which groups them according to whether they appear evolved or planned, and whether they appear to be the product of forces acting comprehensively on the entire township or on selected holdings or sectors. But no matter how we seek to classify it, such variety makes demands on any simple interpretation. Rather does it make the co-existence of a number of causes a virtual necessity.

Yet another dimension through which we can purposefully elaborate the problem is by examining the procedure involved where townships appear to have been planned, in part or in toto. Their order signifies the use of formal schemes of land division. I do not think it is sufficiently recognised how complex such schemes could be. Altogether, we can isolate four distinct questions. First, there is the question of cause (shareholding, partible inheritance, township re-organisation?) Secondly, there is the vital question of what principles were used to guide the division. Landholders naturally sought to maximise some factors at the expense of others. They could maximise the efficiency of farm structure or the freedom over cropping by opting for a system of consolidated holdings. Or they could set out to secure greater subsistence reliability by trying to maximise the equality of extent and value or the spread of risk through a subdivided field system. Thirdly, they had to decide on the method to be employed in allocating land. Broadly speaking, two methods were commonly used. Either landholders could draw lots or they could rely on the ways shares were designated to fix their order in any sequence of allocation. A fourth question can be entered regarding the various devices or means by which these different methods were implemented. Thus, the use of a lottery could be based on symbols to identify the person (personal tokens) and his place in each sequence of allocation, or on symbols to identify the holding (stones, twigs, turf slices) as they had been divided out on the ground. Systems that relied on the more specific designation of shares included that of sun division (sunny and shadow shares), numerical ordering systems (first, second and third strip, etc., throughout the township), house or toft sequence (as in some regular villages) and those based on the possible contributions to a joint plough-team (as in the Welsh law codes). The laying out of a regular subdivided field system, then, could be a complicated or staged process (Dodgshon, 1975d).[6] Too often, interpretations have centred on only one stage, without realising the need for specifying what was going on at other stages. Here again, there is scope for some mixing of ideas.

Subdivided Fields: A Structured Interpretation

While such proposals have the broad explanatory power needed to encompass the variety of subdivided fields, and the different processes associated with them, it hardly removes the lingering doubts over why such a highly individual institution should reflect the convergent effects of different causes. But even this obstacle can be negotiated when it is appreciated that our apparently independent processes do, in fact, stand in some relationship to each other. Indeed, when this relationship is depicted, it prepares the ground for constructing a single, compound interpretation that covers not just the different forms but the changing circumstances of subdivided field formation.

One way of sorting out their hidden relationship is to ask whether the interpretations discussed earlier are really as independent as past reviews have made them out to be. Mention has already been made of the suggestion that partible inheritance can be seen as a particular instance of Vinogradoff's shareholding interpretation, one of a number of ways by which townships could become shared. Indeed, it must be construed as such for it to have the ability to intermix as opposed to merely dividing land. Likewise, McCloskey's risk aversion ideas can be fitted into a shareholding interpretation as part of the guiding principles behind division. To some extent, the most awkward interpretation to handle in this context is that based on joint ploughing. But while it is seemingly entrenched both in literature and song as a cause, it still lacks proof as such. To accept these conclusions means that the genuine options before us can be reduced to two basic interpretations: that of shareholding and that of piecemeal colonisation.

What makes these two so appealing is that they cover the extreme forms of subdivided field layout: the one able to generate evolved forms and the other, planned ones. At the same time, both were able to work selectively (on holdings or sectors) or comprehensively on the system. Still more attractive is that whereas they cannot be fused into a single hypothesis, they can be linked together in compound fashion by seeing the one as making out the case for the other. In theory, both could conceivably act as precursor for the other: shareholding could foster the systematic sharing of land during phases of expansion just as easily as the latter could engender a spirit of shareholding. Any attempt to build a compound interpretation, therefore, must answer the question — which came first?

I would like to frame this question in the following way. Subdivided fields represent a special kind of relationship between pairs or groups of landholders, a relationship based not merely on adjacency but on interjacency of their property. It was a relationship between landholders which, by the mere interweave of property, imbued them with a common or overlapping sense of territoriality. What we are asking is how they first got into this relationship. I have couched it in these terms because I feel it represents a more neutral approach than that which talks from the *very beginning* of open-field communities as if they were functional and behavioural groups. The latter carries connotations of them being organic communities, or born that way, when, in reality, they may have been held together by a mechanical relationship.

A tempting route would be to follow Vinogradoff in seeing shareholding as a form of modified tribal tenure (Vinogradoff, 1892, 127-8; 1905, 52-3; 1920, 321-43. Needless to add, this effectively endows shareholding with a degree of archaism that would make its

operation as a primordial influence on subdivided fields a reasonable
supposition. However, it carries with it other implications. For instance,
it means that subdivided fields were underpinned by a communal tenure.
According to Vinogradoff, they never lost this link with communal
tenure so long as they remained unenclosed (Vinogradoff, 1920, 340).
To agree with this means that we must re-phrase the question I posed a
moment ago about how landholders got into the special relationship we
call subdivided fields. The question would now have to become how
they got out of it for the interdependence of tenure and holdings in
some form would now become the starting point. We might combine
the ideas of Vinogradoff with those recently expressed by D.C. North
and R.P. Thomas by seeing subdivided fields as the on-the-ground
expression of the decomposition of common property forms to
exclusive communal forms (North and Thomas, 1977). Logically, to
establish the close association of subdivided fields and communal
tenure, and to give the township a tenurial status of its own, Maitland's
personna ficta[7] (Maitland, 1911), makes it easy to understand how
piecemeal colonisation worked quickly and effectively to extend the
system.

But the weakness of such a view is that it takes too much for granted.
In the first place, it is not proven that subdivided fields can be linked
with a communal tenure. Even their most primitive form, or Scottish
runrig, is more realistically viewed as an example of several tenure
(Dodgshon, 1975a; D'Olivier Farran, 1959). Another reason for
questioning whether shareholding was the starting point stems from
recent work which stresses the importance of several or consolidated
holdings as a precondition of subdivided field systems. Hoffman, in
particular, emphasises this point (Hoffman, 1975, 54). If accepted, it
means that our initial interpretation must contend with a nucleus of
consolidated holdings. Finally, a shareholding interpretation would
cause difficulties when dealing with the discontinuity that must exist
between evolved and planned systems. Some of the former were
probably planned or regular systems that had been subjected to
piecemeal amalgamations and enclosure. However, others owed their
character to the fact they had grown by piecemeal colonisation.
Admittedly, shareholding may have assisted in this growth when new
sectors were being colonised, but it needed only a very elemental form
of shareholding for it to produce intermixture, not the full-blooded
tenurial system which Vinogradoff had in mind.

To invert this approach and to see piecemeal colonisation as the first cause of intermixture between holdings has the merit of meeting these criticisms more satisfactorily. It can act on a system of holdings that, in the beginning, were several by tenure and consolidated by nature, requiring only that the surrounding waste should be a common waste whose colonisation was a matter of common interest. The free use of the word common here must not be allowed to deceive, for what was at stake was the conversion of common land to several use *in respect of several rights*. But as with shareholding, there are obstacles to be surmounted when trying to relate piecemeal colonisation to the basic distinction between evolved and planned systems of layout. What I propose, and what is the very crux of my argument, is that this distinction is fundamental to the question of how our two interpretations can be linked to form a compound interpretation. Given the conditions which I have just outlined, piecemeal colonisation can fragment landholding. It may not do so in an orderly fashion but that is not the point. All that matters is that interjacency is produced. If a holding within that system (through partible inheritance?) or if the entire system (through the adoption of a two/three field system or change in tenurial structure?) was re-organised (Thirsk, 1964, 3-29; Sheppard, 1976; Wade-Martins, 1975; Roberts, 1977, a; Dodgshon, 1978), it would surely focus attention on what was considered equitable for each holding or unit (bovate, virgate, etc.) in the new layout. Under the circumstances that had emerged, landholders might reasonably have wished to map back into landholding the same notions of what was fair and just that had grown up with the unfolding character of pre-existing layouts. In this way, the idea of shareholding as a stricter definition of tenurial right — equalising both extent and value for each measure of interest — may have been something into which landholders opted at a fairly late date.

But there may have been more to this critical shift than the need to partition a family holding or to re-organise the township. Such forces may have been operating for some time before shareholding as a concept became relevant. Time may have been required for expansion and holding fragmentation to prepare the case for such a view of landholding. The socio-economic context of communities may also have played a part. So long as land was abundant, then property rights may have remained loosely defined. It may not have mattered if the partition of patrimonies did not display a fastidious concern for equality when waste still beckoned would-be colonisers. Indeed, there is even the possibility that such partitions could have worked against the dispersal effects of piecemeal colonisation at this point by trying to localise what the latter dispersed. But over time, arable would have expanded at the expense of waste. Additionally, the progressive fission of family holdings would have worked to reduce their size per family. As the boundaries of one expanded outwards, a situation would have been reached when they threatened to meet equally critical boundaries contracting inwards. This pincer movement could have formed the incentive for adopting a stricter view of property rights when the occasion arose to divide or re-organise property. Only then might they have taken up the logic of holding layout and injected their definition of landholding with notions that each unit of landholding or measure had equal right to the different types of land within the township. The relationship which subdivided fields express, therefore, possibly evolved out of circumstances before being cemented by choice. It was not necessarily one which farming communities were somehow born to.

Maitland once said that he felt we hurried the history of our towns and villages too much, and that there were some ideas that did not come to men until they were crowded close together (Maitland, 1898). His advice needs to be heeded when dealing with subdivided fields. Far from being a lesson in how *primitive* notions of communalism gave way before the disruptive influences of individualism and severalty, their history may have a different story to tell. Over the early medieval period, if not before, patterns of discrete or several holdings may have expanded and, in so doing, stumbled their way into a system of fragmented and intermixed property rights. The mantle of communalism, which regular layouts and open field husbandries testify to, may have been donned later out of choice (or lordly coercion) when the opportunity arose to re-organise or rationalise the system. Even so, this communalism was only a synthetic relationship, propelled along by several interests and tenure.

Notes

1.   Precision Tapes Ltd, ZCCAS 1080.
2.   For instance, in Scotland, there is ample evidence in Barony Court proceedings showing that whilst joint ploughing was the norm, nevertheless, holdings were laid out by birlawmen and related to the tenure of land not co-aration. Even the process by which these holdings were then allocated to the various landholders was unrelated to co-aration. Illustrative of the abundant evidence for joint ploughing is Thomson (ed.) (1919), which documents the case of two tenants who shared a plough-team but who had fallen out with each other. Evidence for the laying out of runrig as a separate process can be found in

Dodgshòn (1980).

3.   Baroness Gray v. Richardson and others, pp. 1031-78 in *Session Cases*, 4th series, III (1875-6), especially pp. 1047-50.

4.   For instances of support, see G.R.J. Jones in Baker and Butlin (1973) and J. Storer Clouston (1918). But the absence of partible inheritance as a custom from much of Highland Scotland by the end of the medieval period is brought out in A.J.G. Mackay (1897-8) and by much recent work.

5.   The best case-study of the interaction between piecemeal colonisation and partible inheritance is A. Baker's study of Gillingham manor in Baker and Butlin (1973), 393-401.

6.   The writer has tried to bring out this potential variety of method and procedure in his 'Law and Landscape in Early Scotland' (1980), but some comment can also be found in R.A. Dodgshon (1975e), 1-14.

7.   It should be added that Maitland (1911), 13 coined this phrase in order to criticise the concept that lay behind it.

7 TOWNFIELD ORIGINS : THE CASE OF COCKFIELD, COUNTY DURHAM

Brian Roberts

The origins of open, common subdivided field systems antedate detailed documentation and as it is virtually impossible to discover proprietary arrangements and rotational practices by archaeological means, much of our understanding of their genesis must depend upon the evaluation of probabilities and possibilities using hypotheses derived from known facts and assumptions. Any study of a field system or village demands information upon four aspects: the form elements of which it was comprised; the way these were integrated into the functional patterns of landholding and farming practice; the forces, internal and external, creating change or equilibrium, and finally the distribution of each definable system or type within space and time. The question of definition is a troubled one. Definition and classification are necessary for the purposes of study and comparison, and Bruce Campbell's paper in this present volume is an important step forward. Such a check list of variables, in spite of undoubted practical difficulties, permits evidence, however full or inadequate, to be rigorously evaluated: what can be proven and what must remain assumption are made clear. The author must admit to less rigorous standards of definition in this essay: the presence of open, common *or* subdivided fields (giving each word equal weight but precise and different meanings) is a proper topic for scholarly debate. Rather than enter such troubled waters it is proposed to use the word 'townfields' as a concise way of describing those fields — divided into strips, with fragmented ownership and discrete holdings — which were often, probably normally, farmed in common and which surrounded many rural villages and hamlets of lowland Britain before the major enclosure movements occurred. They were after all the fields of the 'town', or village (Baker and Butlin, 1973, 619-56).

Townfield Origins

Figure 15 provides focus for a number of points concerning townfield origins. It suggests that the ultimate origins of those variables combined within an open, common subdivided field system of the Midland or

Champion type must be found in two sources, prehistoric and Romano-British systems and those arrangements brought or developed by Anglo-Saxon and Scandinavian folk or Norman lords. The character of such *antecedent systems*, which presumably surrounded not only excavated pre-Conquest settlements such as Charlton, Hampshire, Thirlings, Northumberland and Catholme, Staffordshire, but also many hundreds of settlements whose place-names attest pre-Conquest settlement, remains largely an open question, illuminated only by tantalising glimpses provided by the evidence of charters and laws — 'If ceorls have common meadow or other lands divided into shares' etc. (Ine, 690 AD). The question marks appended to the upper stippled plane (Figure 15) indicate these uncertainties but imply that the division between Champion townfield systems and other systems is indeed worth discussion and consideration (Loyn, 1962; Roberts, 1973, 194).

Approaching the problem retrogressively, the known distribution of townfields in Britain and Europe makes it clear that there are areas where they dominate and areas where they are less apparent (Plane A of Figure 15). The term 'favourable' begs many questions, for the environment may be social and economic as well as the conditions of terrain, soil and climate. 'Favourable' physical environments in no way 'produce' fully fledged townfield systems. However, the presence of zones once utterly dominated by such fields, for instance the Feldon of south Warwickshire and the valleys and plateaux of east and south-east Durham, in close juxtaposition with other zones where townfields had a more limited distribution and where enclosures in severalty are dominant, the Arden of north-western Warwickshire and the Pennine foothills and dales of western Durham, pose a particularly significant question: are such boundaries the result of contrasts in physical conditions, a response to varied land qualities, or must they be seen in the context of a process of townfield development which for some reason was arrested? Can the irregular field systems of Arden and the irregular systems of western Durham be considered as antecedent forms, such as might if circumstances had allowed (and here the physical environment must be seen as one constraint) have developed into fully fledged townfields? Both zones were regions where active colonisation continued into the twelfth and thirteenth centuries, where townfield land remained a very low proportion of the total area of each township, and where enclosures in severalty offered a wide range of economic and social opportunities. Significantly, in Arden the author was only able to prove townfield expansion during the thirteenth century in a township well developed by 1086, peripheral to the main late settled zone where

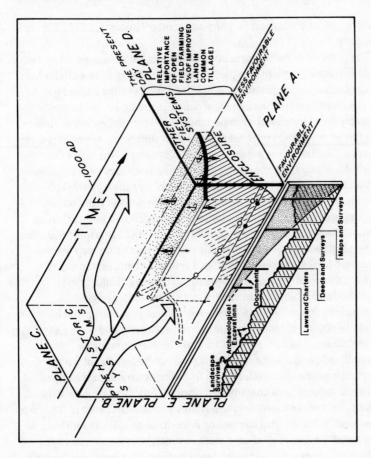

Figure 15: 'Open' Field Systems: Development and Documentation

the strength of communal farming was greater (Roberts, 1973, 226-9; Hodgson, 1979).

The progress of enclosure in boundary zones between contrasting areas often involved the silent and piecemeal restructuring of townfields long before the period of classic open-field enclosures of the second half of the eighteenth century, although, of course, Durham lowland townfields are exceptional because they experienced seventeenth-century enclosure. It is fair to ask, however, if there was ever a time when townfields were in fact advancing at the expense of earlier systems — the arrows in Figure 15 penetrating the stippled plane (the boundary between areas dominated by townfields and areas dominated by other systems) suggests possible trends. The comparison of field forms derived from widely differing spatial and temporal contexts is an attractive but dangerous exercise; thus, the layout of Cockfield, County Durham, to be discussed below (Figure 16) resembles in some respects that of the surviving open-field (*sic*) clachan of Ballintoy, County Antrim, and in other respects the Eschfluren of north-western Germany (Uhlig, 1961; Mayhew, 1973). Similar forms need not mean similar origins, but, nevertheless, forms found in close geographical proximity, which appear to reveal gradations in dimensions, regularity and degree of communal organisation, warrant careful consideration, more especially when that same transition zone can be associated with temporal contrasts in development. Such forms may indeed be the legitimate subject of taxonomy and possibly reflect an evolutionary process. To assume such evolution is, however, not to prove it. The date when townfields stopped expanding is a critical one, for from then onwards all processes of change were internal, even when related, by farming practices for instance, to lands outside. Of course, no certain, universal or spatially coherent *terminus post quem* is possible, and as late as the early seventeenth century newly added furlong units, subdivided into strips on the basis of house-order in the village, were added to the townfields of Acklington, Northumberland. Nevertheless, it is worth reflecting upon what the cessation of townfield expansion involved, upon some possible causal linkages and a possible chronology (Roberts, 1978).

In physical terms cessation of expansion meant no furlongs were added, either by the addition of whole new furlongs, appropriately subdivided into parcels, or by the more gradual process of absorbing lands cleared in severalty. Such a physical limit, it must be stressed, in no way excludes developments internal to the townfields: further subdivision, new rules of cropping and grazing and the reorganisation of

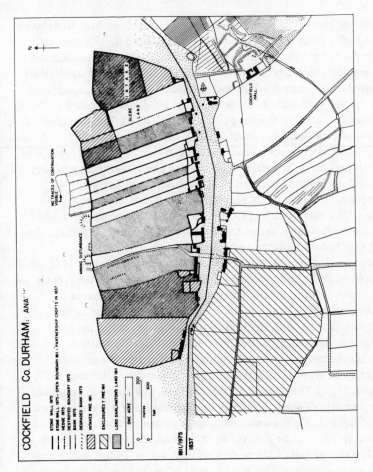

Figure 16: Cockfield Co. Durham: Analysis of Plan

holdings. There is, however, one important variable omitted from Bruce Campbell's check list, the presence of standardised fiscal tenements, oxgangs and bovates, virgates and yardlands, the former clearly based on ploughing capacity, the latter clearly involving *measurement,* albeit derived from the ox-goad. It may be easier to fix a *terminus post quem* for the non-inclusion of land within what Dodgshon has called the 'land unit system' of each township (Dodgshon, 1975, c). The pre-Conquest origin of such fiscal tenements is hardly in doubt — a point of paramount importance when considering the origins of field systems of which they are an integral and functioning part — and these represent the full status holdings to which Sally Harvey's bordars were peripheral, not only in the geographical sense, but also in social and economic terms (Jolliffe, 1935-6; Sawyer, 1976). The author would not care to provide a *terminus post quem* for such tenements in Warwickshire but the Durham evidence, to be discussed below, is clear and unequivocal: the bovated lands described in Bolden Book of 1183 are by and large the bovated lands of the Hatfield Survey of 1381, lands reclaimed between the two dates rendering merely a cash rent — exchequer lands. Constance Frazer dated a similar change on the Priory estates to about 1200, and while non-bovated townfields are not unknown, the author has documented elsewhere how Durham townfield acreages expressed in bovates of the same size remained constant between 1183 and the seventeenth century (Roberts, 1977, b; Fraser, 1955). The townfield-fiscal tenement link seems an important one, the date of about 1200 may be significant, and it is reasonable to ask, for within the present context more cannot be done, what broad causal factors may be involved in establishing this temporal datum?

The second half of the twelfth century, a period when justice and finance 'were inextricably bound together', saw a number of important changes in the national scene. It is to changes in justice, taxation, society and economy that we must look to find those powerful forces capable of generating changes in both field systems and settlement. The rise of the real actions, not least the assize of Novel Disseisin which provided freeholders with a speedy remedy against ejection must have embraced appended and appurtenant rights over the common waste, and it was not until the Commons Act of 1236 (the Statute of Merton) that what seems to have been a practical rule of common law allowing the lord to fence, enclose or approve any land of the waste as against his freehold tenants holding rights of pasture was formally placed on the statute books. How far, in the latter part of the twelfth and the early thirteenth century the availability of Novel Disseisin acted as a

constraint upon approvement is uncertain; its impact must have
depended upon the social structure of each manor and the extent of the
local manorial wastes. Novel Disseisin was probably instituted by the
Assize of Clarendon (1166), while the year 1162 saw the commencement
of an important financial change: Danegeld was levied for the last time,
and while in 1194 it was again taken, it was under the name 'carucage'
and subsequent levies in 1198 and 1200, no longer based on Domesday
Book, were assessed by special commissioners and comprised a land-tax
levied from the carucate of ploughland. There were, however, important
late twelfth century experiments with taxes upon moveable goods. This
same period also saw a tendency for demesne farms to disintegrate, so
reducing the demand for labour services; indeed the leasing out of
portions of demesne land to the tenants themselves was often
accompanied by forceful pressures from the tenants to free themselves
of this type of rent obligation in favour of money rent. By the 1190s
this trend reversed, with landlords gradually resuming that active
cultivation of their demesnes, dependent upon labour services, which is
such a feature of thirteenth century economic life (Simpson, 1961;
Postan, 1972; Poole, 1955; Roberts, 1975). To balance these forces, to
explain their detailed action at a local level is difficult, but it appears
unassailable that this is the context in which new fiscal tenements
ceased, in general, to be added to townfields, and a broad *terminus post
quem* is to be sought between the middle of the twelfth and earlier
decades of the thirteenth century.

This argument clarifies our problem of origins: on the one hand we
have to trace those developments continuing within townfields following
the cessation of expansion, and on the other there remains the question
of the developments prior to this date. One point should be stressed:
open, common *or* subdivided fields could exist before the emergence of
those systems which integrated significant numbers of Campbell's
diagnostic variables. Thus, many of the individual elements —
differential gears, pneumatic rubber tyres, clutches and petrol engines —
were already in existence before 1895, when they were integrated to
create the antecedent of the modern motor car! The adoption of the
integrated set of variables present within fully fledged open-field
systems need not, of course, have been spatially rational, primarily the
result of horizontal diffusion. A strong measure of hierarchical diffusion
was probably involved: the result of the seigniorial imposition of new
arrangements upon certain groups of settlements, for instance, those
with a higher proportion of servile tenants, those containing a demesne
farm or those in need of revitalisation as a productive estate. These final

points carry the argument back to Figure 15: take two field systems, documentable, within the more recent section of plane E. These are shown in plane B as two dots, one open, one closed, drawn close together. If these individuals could be traced backwards in time and their trajectories and antecedents traced, then it is possible that they would appear to diverge, the two similar systems of later centuries having different roots. To put this another way, the rise of townfields could be associated with a process of convergence, with successions of small changes, the result of population trends (be these rising or falling), inheritance practices, new agrarian technology, improved rotations, variations in yield or the degree of seigniorial control, combining to accrete to an antecedent system those formal and functional structures characteristic of open, common, subdivided fields. Such a model, although infinitely more complex, would put into a wider perspective those small differences between adjacent field systems which are apt to be ignored by generalisation.

The Case of Cockfield, County Durham

The remainder of this paper is concerned with one possible antecedent type. Hodgson and Butlin have outlined the key characteristics of the field systems of County Durham: a critical landscape divide runs from the Tyne valley to Chester-le-Street, Durham, Bishop Auckland and Barnard Castle (Hodgson, 1979; Baker and Butlin, 1973). To the east lay villages supported, until the seventeenth century when enclosure occurred, by communal townfields. It is unlikely that even within the Tees valley or on the Magnesian Limestone these ever represented more than about 70 per cent of the total township area, and in addition to extensive areas of lowland moor and fell there must have often been amounts of older enclosures. To the west, however, the proportion of townfields decline dramatically, with fells and old enclosures increasing in importance. The distinction between the two zones, and the implications for field systems is indicated by a comparison of Boldon Book (1183), the Hatfield Survey (1380-1) and the seventeenth-century Chancery Decree Awards concerning enclosures, mapped by Hodgson. In 1183 the vills held by the Bishop in demesne (but not necessarily containing demesne land *per se*) are bovated, each holding normally being of two bovates. There are technical problems here: Durham men traditionally never answered for the geld, i.e. never paid Danegeld, and as Jolliffe points out these double bovates probably conceal 'husbandlands'

(Jolliffe, 1935-6, 162). Nevertheless, these seem to have been townfield holdings whatever the minutiae of their tenural implications. These holdings are normally described in 1380-1 as *Terrae Bondorum* and there can often be a direct relationship between the bovated land, not only in 1183 and 1381, but with the number of assessed holdings in the early seventeenth century. The distribution of exchequer land, land measured in acres and rendering a cash rent only, is particularly revealing: it is almost entirely concentrated west of the key divide, those amounts recorded in east Durham being very limited in area and significantly highly subdivided into small fragments. Exchequer land clearly represents that land reclaimed *after* 1183, and the amounts involved can, in the west, often total hundreds of acres (Greenwell, 1857). It is quite clear that, as argued above, at some stage a distinction emerged between land seen in terms of a standard land unit and land newly added to the improved area. As active colonisation continued, in spite of increasing problems of terrain and climate, and as in Weardale cultivation (of wheat?) extended to 1,100 feet (335m), spreading permanently improved fields to a general altitude of 800 feet (244m), there appears to be no sound grounds for explaining the presence or absence of extensive townfields within some foothill villages entirely in physical terms (Birks and West, 1973, 207-21). At Hamsterley, at 600 feet (183m) in the foothills townfields were not only present to be recorded on nineteenth-century maps but survived to be remembered in the twentieth century, and indeed in some senses, remain visible today, albeit consolidated in the hands of one man. Cockfield, the subject of a case study here, lies in the same type of terrain, at about 700 feet (213m): it retains to this day a visible field system, which may represent one antecedent type, although one must hesitate to apply the term 'townfield' to it.

Cockfield is not well documented: the earliest reference in the 1220s, is a personal name within a charter, and the manor was passed by the Vavasours, without the Bishop of Durham's permission, to Ralph Earl of Westmorland in about 1411 (Newcastle upon Tyne Records Committee, 1929; Deputy Keeper of Public Records, 1883-4). This raises many questions concerning overlordship: Cockfield was part of either Aucklandshire or Staindropshire and the interest of the Bishop suggests that it may have been attached to that section of Aucklandshire, south of the Gaunless, acquired from Staindropshire during the tenth century. Nevertheless, if the Vavasours had been holding it of the Bishop then it ought to appear in the Hatfield Survey. It does not: an omission rendered more surprising in that an inquisition *post mortem*

of 1349-50 shows the family holding of the Bishop, and Cockfield
appears to have been in the hands of a knight of the Bishopric between
1264 and 1272. Certainly the estate after 1411 passed via the Earls of
Westmorland and the Vanes to the Dukes of Cleveland and so to the
present Lord Barnard (Greenwell, 1857). Since 1411 it must, as part of
these estates, have looked southwards towards the Tees valley. The
place-name, incorporating a personal element and 'field', implying
'open country, land free from wood', implies possible Anglo-Saxon
origins for the settlement, but as it is not mentioned in the *Historia
de Sancto Cuthberto* of 1050, and as a church, much of whose visible
fabric is Early English, was present by 1291, then development between
1050 and that date can be inferred, development which must in fact
antedate the earlier decades of the thirteenth century. The signs are,
with the absence from Boldon Book, that as late as 1183 it was still
relatively underdeveloped, although Evenwood, referred to in the
Historia is also absent in 1183 (Greenwell, 1872, 138-52).

The present village is a straggling two-row plan with an irregular
street green nearly a kilometre in length. Figure 16 summarises the
evidence of nineteenth century map sources while Plate 2, the air
photograph, allows the reader to assess some of the evidence as the
argument proceeds. The most striking feature of the plan is a series of
long-tofts on the north side of the settlement, on a north-facing slope
extending across a drift-covered sandstone shelf, a singularly unpleasant
location, facing the north and east and being exposed to bitter winter
and spring winds (Smith, 1970, 58-74). The remainder of this slope
is unenclosed bracken and grass fell. The outer perimeter wall of these
long enclosures is of indeterminate date and heterogeneous in
character, with each proprietor enclosing his parcel-end, and usually
leaving a gate. It is, however, comprised of field stones, with rounded
angles, or very ancient weathered quarry-stones, rather than the fresh
angular quarry-stones normally associated with classic late-eighteenth
and nineteenth-century walling. It lacks the projecting 'throughs',
through-stones, giving two lines of stonework standing clear of the wall-
face which are another diagnostic feature of 'recent' walling.
Significantly, the boundary walls between the long-toft parcels are
mainly of quarry-stone and bear 'throughs'. There are hints in the
perimeter wall of the presence of rectangular much weathered quarry
blocks approximately 30cm $\times$ 20cm $\times$ 20cm.

Close study reveals three things: near to the perimeter wall and
parallel with it are traces of some discontinuous low earthen banks,
presumably replaced by the stone structure; beyond these, extending

Plate 2: Aerial Photograph of Cockfield, Co. Durham

Source: Durham County Council and BKS Surveys Ltd.

for a further 50 or 60m out onto the fell are a set of low banks, of earth, but with some stone incorporated, which terminate in a continuous outer dyke, the former edge of improvement. These continue the inner toft boundaries outwards. Thirdly, the present perimeter wall is at times quite irregular, and angular offsets of as much as one metre occur; at one of these junctions the presence of a massive embedded boulder suggests the use of meer-stones. The character of these earlier boundaries cannot be proven beyond doubt but what is clear is that the visible stone-walls at Cockfield are superimposed upon an older set of boundaries, some clearly linked with ditches, others merely comprising low banks with no trace of an attendant ditch. A careful examination of the levels between each long-toft shows that were they stripped of all upstanding boundaries then the North of England would acquire a fine set of strip-lynchets with risers of the order of one metre, more rarely a metre and a half! A close study of Figure 16, removing as it were the later peripheral intakes, shows that the earlier earthworks encase a very regular plan, with a group of shorter tofts at the eastern end of this north row, longer tofts further west, with uncertainty concerning the western end of the plan. The toft-head line, along the street, shows traces of two *en-echelon* building lines, the eastern lying somewhat south of the western.

How are these observations to be interpreted? The stone-walls are evidently replacement boundaries, but an absolute chronology is difficult to establish. An estate map in Lord Barnard's collection, dated 1811, shows some of the long-toft boundaries as pecked lines, a group called 'Partnership Crofts' on the Tithe map: presumably they were once open parcels, and the same maps reveal fragmented ownership patterns. The location of the glebe is interesting; it was surely present by 1291 and gives a *terminus ante quem* for the northern row, the western end conceivably being somewhat later in the development sequence. Identical long-tofts at Iveston and Frosterley and other foothill villages suggest that this distinctive plan-form is to be linked with settlements undergoing development during the latter part of the twelfth and early thirteenth centuries, a hypothesis confirmed by the case of Byers Green, merely an assart in 1183, but probably of village status by the earlier decades of the thirteenth century (Roberts, 1972). On balance a *terminus ante quem* of as early as 1200 may not be inappropriate for this type of layout, although conclusive proof is difficult to obtain.

To the south-east of this northern row, to the south of and below the church, lies the moated Cockfield Hall, and this permits a basically simple hypothesis to be formulated:

1. The original settlement lay near the hall, on a more sheltered freely-draining site. A colliery spoil heap to the east, a nineteenth-century steam-engine pond in the middle, universal subsidence and heavy poaching by stock and vehicles give a lie to this interpretation, but sandy drift is quite evident in exposures. The surrounding earthworks, with the exception of the moat, are not medieval, largely being the result of ploughing, during the last war, the nineteenth century and earlier, together with at least two phases of field drains. The site, however, is an interesting one and futher evidence for its importance will be considered below.

2. The expansion of the settlement in the twelfth and early thirteenth century, probably linked with the expansion of the chief farm to create a larger demesne, led to the foundation of the church and the creation of a new tenant row along the ridge to the north. The long-tofts, like those at Wharram Percy, represent a foothold furlong for new or displaced tenants, may have been intensively manured and were small enough to have been intensively tilled by one family. Measurement of their frontages strongly hints at the use of a 20- or 21-foot land-rod to lay them out, the latter being the so-called Durham rod.

This hypothesis begs the question, where were the original lands of Cockfield? The air photograph reveals a second field area, to the west of the hall, oval in shape, but at its western end bearing traces of a pattern of plough-curved ownership parcels. The north-western boundary, 'Watery Lane', is followed by a stream, and consists of a hollow way, some two metres below the general level of the field land to the south, from which it is separated by a near-vertical slope, this being penetrated by clear access ramps, one for each hedged parcel. The southern boundary is also followed by a water-course. The south-east is a well-marked lynchet one and a half to two metres in height, the oval field being the higher, while the south-west has a substantial but eroded bank, one section of which is clearly visible on the aerial photograph as a slight earthwork — indeed it is this archaeological feature which reinforces the distinctive oval shape. The internal enclosures are hedges, placed upon earthen banks, sometimes revetted with stones, sometimes with ditches, but the removal of upstanding structures would again reveal traces of well-marked lynchets, with risers of about one metre. At one point on the southern boundary a massive erratic block may represent a meer-stone – it lies exactly at the southern end of one cross-boundary! This oval area forms a low, south-canted, rounded dome, well-drained, south-facing, probably with deeper soils than are found on the northern slope. The manor house is clearly associated with its eastern end and all but one of the northern access ramps point to the east. The attractiveness of the micro-climate of this field area is revealed on any day with the wind in the northern sector! The author would stress the distinctive qualities of this 'field', for such it must be. Personal observation, and work by Mr Huw Evans shows that in 1838 its western half was divided into six individually owned strips, although today a measure of consolidation has occurred, with council housing invading one parcel.

These observations permit an alternative hypothesis to be formulated and Figure 17 attempts a diagramatic reconstruction of the events at Cockfield, together with their likely chronology:

1. Field I, the southern oval, is an early focus of agricultural activity and probably represents the arable nucleus from which later arrangements sprang. Settlement at this stage lay near or on the manor house site. The scale should be grasped: about 25 acres (10 hectares) is involved. It resembles those carefully manured infield cores or kernals described on a European scale by Uhlig (Uhlig, 1961). It is hoped that phosphate analysis may support the hypothesis that its soils have been dunged and husbanded for centuries (Alexander and Roberts, 1978). A division into strip holdings and perhaps controlled grazing of the stubbles must be postulated.
2. Field II, the northern row, represents a secondary expansion, earliest in the east, with subsequent accretions along the ridge to the west.

   The boundary of the farm associated with the manor house extends to the roadway dividing Field I into two, thus permitting the attractive suggestion that demesne expansion, taking over a large share of the old field, necessitated the addition of the extra farms to the north, and this would be represented by the shorter tofts at the eastern end of the northern row, including the glebe of course, which has no share in the old field although it does, significantly, have more land to the west (Figure 17). This created new house plots north of the church, and a slightly later expansion gave the main series of long-tofts, with their associated houses, and completed the shift northwards of the settlement focus.

The arable holdings would, initially, have been very small and if the

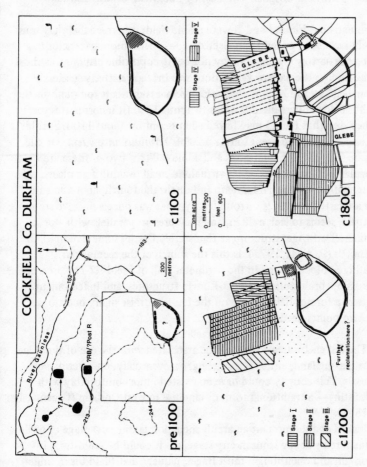

Figure 17: Cockfield: Evolution of the Field System

argument that Fields I and II were arable fields is correct their hay must surely have been culled from elsewhere — other temporary enclosures along the narrow Gaunless valley not being impossible although as the name implies the narrow valley floor was indeed relatively 'useless' (Dewdney, 1970, 261). A chronology remains difficult to establish: the author believes that Field II dates to about 1200 (it incorporates the glebe), and that Field I may have been present by about 1100, but this latter date must represent only a possible *terminus ante quem*. On the fell are four ancient enclosures and a linear ditch: two of these are irregular rectangles, and one is virtually a small oval hill fort placed on a slope, with a substantial rampart and ditch; the fourth, about an acre and a quarter in area, is surrounded by what was once a massive stone rampart, seems to lack a ditch, and may present parallels with a post-Roman, 'Dark Age' structure at Hamsterley Castle, some five miles to the north (Roberts, 1975). Is this the source of the rectangular weathered blocks noted in the perimeter wall of Field II? The presence of possible prehistoric and post-Roman structures, and indeed slight hints of a former field system on the fell must raise many questions, but two points need emphasis;

1. The fell enclosures could well be associated with the use of the ridge as grazing land, either as a shieling, or, eventually, on a permanent basis. This activity could have co-existed quite comfortably with hunting — a traditional form of land use in west Durham (Greenwell, 1852).
2. Field I could represent an arable incursion taking advantage of well-manured shieling lands at any stage, i.e. it could be prehistoric in origin, and one thought must linger, namely that Field I is in origin merely a larger prehistoric enclosure. Field II could also be linked to the introduction of a two-course shift — its area is of the order of 25 acres (10 hectares), but is difficult to calculate because of the problem of defining the western boundary (Slicher van Bath, 1963).

To return to the theme of townfield origins: the field morphology of Cockfield suggests the presence of two old cores, probably originally containing open strips, under subdivided ownership in the nineteenth century, and with the older of the two having high value as potential arable land (Green, Haselgrove and Spriggs, 1978; Roberts, 1978). No evidence survives to speak of functioning arrangements. Striking parallels for this type of structure can be found in Sweden, particularly in the irregular settlements which preceded the regular *solskifte* arrangements, and a by-law, from the village of Sunnanå in Lillhärdal in Harjedalen cited by Erixon, will raise many echoes. It was registered in 1777; 'anyone who has fenced in private estate, whether one or several, that is to say, anyone who in the village itself or outside it has enclosed field or meadow, must keep a wattled fence valid in law around this or blame himself. If any damage is done on such enclosed estate no other person shall be fined or penalised. Those who have gathered about such an enclosure may leave their cattle to graze in autumn or spring' (Erixon, 1966). Of course they could! The weedy arable margins and the stubble after harvest provided feed to fatten in the autumn and to give that vital early bite which can spell success or failure in carrying beasts through the winter. In Cockfield we surely see a similar antecedent form from which in more suitable environments fully-fledged townfields could develop.

Why did townfields *not* develop in Cockfield? To this day the area is marginal for arable; it lies upon a frontier, and must reflect minor climatic variations. It was, furthermore, marginal to the great estates in the hands of great lords, for whom it was peripheral to vastly richer acres elsewhere. Their *tenants in capite* appear not to have had extensive lands, but pastoral production, quarrying and even coal mining must always have been attractive alternatives to marginal arable production from an early stage (Deputy Keeper of P.R., 1883-4, 274). The fossil field structures preserved for us in this difficult environment only intensively settled at a stage later than the more attractive lowlands, reveal features long since vanished from those lowlands. The remains, in addition to giving a glimpse of an 'antecedent field structure', provide a useful picture of the origin of one village plan-type, and reveal how in this case the green, an area of inner pasture, evolved, almost casually, from a driftway leading westwards between the settlement and the grazing lands. Cockfield is a very different type of green village, however, to those of the lowlands, with their rigidly formal plans and often highly-organised townfields. It bears clear traces of more primitive arrangements and in its basic structure, a settlement linked to an inner pasture with adjacent enclosed arable lands, possibly with some meadow, and driftways leading out to rough unimproved grazings, we see those elements of a basic model which may assist us in understanding the economic basis of some pre-Conquest settlement nuclei.

8 THE EVOLUTION OF SETTLEMENT AND OPEN-FIELD TOPOGRAPHY IN NORTH ARDEN DOWN TO 1300

Victor Skipp

A general account of the open-field systems of the Forest of Arden was given by Roberts in his 'Field Systems of the West Midlands' (Roberts, 1973, 188-231). In particular he noted the wide variation in their topographical layout and extent from one parish to another and drew attention to the contrast between the nature of agrarian development below and above the 400-foot (122-meter) contour. This paper looks in detail at the open-field systems of five contiguous north Arden parishes which, taken together, lie between 300 and 500 feet (91 and 152 meters) and its purpose is to consider the siting, layout and possible chronological development of these open-field systems in relationship to settlement.

Of the five parishes, it is no doubt significant that the three smallest — Elmdon (1,127 acres, 456 hectares), Sheldon (2,500 acres, 1,012 hectares) and Bickenhill (3,771 acres, 1,526 hectares) — lie below the 400-foot contour; whereas the two largest — Yardley (7,590 acres, 3,072 hectares) and Solihull (11,296 acres, 4,571 hectares) — possess extensive areas of higher land (Figure 18).

Physiographically the country above 400 feet belongs to a subsection of the Birmingham Plateau known as the Solihull or Arden Plateau; that below 400 feet slopes northwards and eastwards to the rivers Cole and Blythe respectively, and as such forms part of the Tame-Cole-Blythe Basin (B.A.A.S. 1950, 4-5).

Although the whole district is floored by Keuper Marl, on the plateau top this is almost entirely overlain by Boulder Clay/undifferentiated drift, and to a lesser extent by glacial sands and gravels (Figure 19).[1] Another extensive area of drift lies crutched between the Cole and Blythe immediately to the south of their confluence, forming Coleshill and Bickenhill heaths. Otherwise Keuper Marl predominates on the Cole-Blythe valley land. Only restricted areas of drift occur: among them several which cap the narrow interfluvial spurs that project out from the plateau, and that descend on either side to northward- and eastward-flowing tributaries of the Cole and Blythe. On this lower land, too, though not on the plateau top, ribbons of alluvium skirt the main

162

Figure 18: The Five Parishes and Arden Plateau

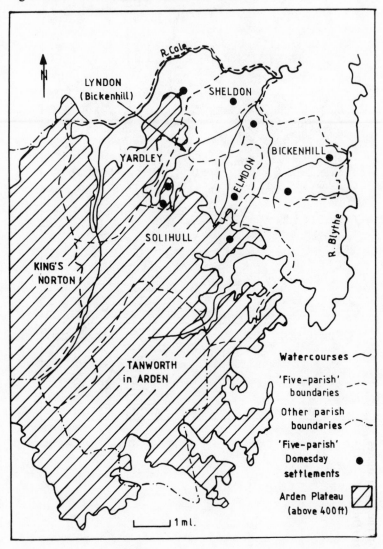

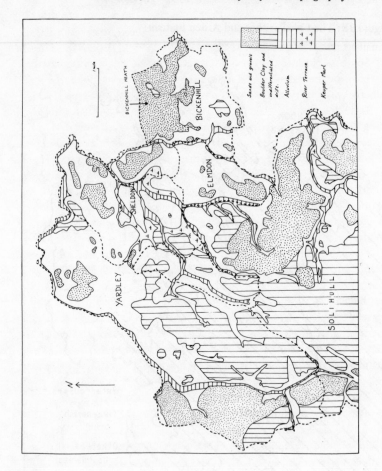

Figure 19: The Five Parishes — Geology (Drift)

rivers and most of the lesser watercourses, and there are occasional small remnants of gravel terraces.

Some insight into the natural vegetation of the area may be obtained by plotting the botanical and clearing names that occur in the mid-nineteenth-century tithe surveys on to a geological map (Skipp, 1960, 6; Skipp and Hastings, 1963, 10). This shows that 'Moor', 'Marsh' and 'Bog' fieldnames were most common on marl (61) and alluvium (44). They are also found on Boulder Clay (28), but are rare on sand and gravel (9). As might be expected, the great majority are adjacent to brooks, but there are several alignments which appear to be related to a junction of drift with marl, and are presumably indicative of seepage from the former. 'Rush' and 'Sedge' names occur on marl (4), alluvium (5) and Boulder Clay (7), but not on sand and gravel; the six featuring 'Alder', 'Withy' and 'Osier' are entirely confined to alluvium.

The 130 'Wood' names (Wood, -hurst, -greave, -shaw, etc.) are fairly evenly distributed over all types of geological formation. But clearing names (Ridding, Stocking, Breach, -ley, etc.) show a heavy concentration on the marl (90), with relatively few on the Boulder Clay (35) or sand/gravel drift (23). Eight of the ten 'Oak' names are also found on the marl, whereas all seven 'Birch' references are associated with Boulder Clay or sand/gravel drift. Finally, 'Broom' and 'Gorse' names are mainly restricted to the lighter formations, with 17 out of 34 on sand and gravel.

The above suggests that within the oak-birch woodland that must have once covered the area, birch may have been the dominant tree on the drift soils, but that the marl carried oak-dominated forest which could only be cleared with the greatest difficulty. Nevertheless the bulk of the marl had apparently been brought into cultivation by 1300, for a comparison of Figures 19 and 22 shows that extensive stretches of waste were confined almost entirely to the drift soils. As Roberts has pointed out, the earliest medieval references to vegetation make a clear distinction between woodland and heath (Roberts, 1968, 102). However, in the parishes here under discussion, wooded and open waste land appears to occur with almost equal frequency on both the Boulder Clay and the sand and gravel, and that fact suggests that the distinction between them is unlikely to have been geologically based; but rather that the heaths of *c.* 1300 probably represented an already devastated form of the lighter oak-birch woodland.

When this process of devastation began it is impossible to say, for evidence of early settlement in the area remains elusive. Aerial survey indicates that the terrace gravels of the lower Tame valley may have been under cultivation from Iron Age times or earlier. But not only are substantial river terraces absent from the five parishes: such inhibiting factors as the widespread urbanisation and the presence of the Elmdon air corridor make it seem unlikely that traces of any Roman or prehistoric field systems which may have existed in other locations will ever be discovered. On the other hand, Berry Mound, an important multi-ramparted Iron Age hill-fort of 11.5 acres (4.7 hectares) lies at Shirley, more or less at the centre of the Arden Plateau; while the finding of pottery sherds of the second to fourth century AD at Maiden's Bower, within a few hundred yards of the Solihull High Street, raises the possibility of some kind of Roman settlement there. Perhaps the balance of opinion still favours the view that throughout prehistoric and Roman times the Birmingham Plateau and its immediately adjacent areas 'remained for the most part in their natural state, forests, light woodlands and scrub', exploited, in so far as they were exploited at all, for little more than 'hunting and pannage and providing wood for charcoal burners' (Webster, 1974, 55). However, the recent surprise discovery of a major Roman villa and/or temple site at Coleshill serves as a reminder that a balance of opinion is by no means the same thing as certain knowledge.

The degree of development at the end of the Saxon period is of course easier to assess. The Domesday survey records nine settlements within the boundaries of the five parishes (Figure 18). And despite the lack of any pre-Saxon evidence from below the 400-foot contour, it is clear that by this time, with six of the Domesday manors in the Tame-Cole-Blythe Basin, it is this lower land that represents the main area of growth. Furthermore, even the three settlements that are to be found above 400 feet are on the very northern edge of the plateau, leaving its extensive hinterland apparently devoid of activity.

The commonfields, meadows and wastes of the five parishes on the eve of eighteenth- to nineteenth-century enclosure are shown on Figure 20. By 1700 only about 3,000 acres (1,214 hectares) of communally managed land was to be found, which amounted to 11.5 per cent of the total acreage. Out of this, 1,250 acres (506 hectares, or 4.7 per cent of the combined parochial areas) was common arable, 148 acres (60 hectares, or 0.6 per cent) common meadow, and 1,600 acres (648 hectares, or 6.1 per cent) common pasture.

However, in the Forest of Arden at least, the extent and form of common land in 1700 cannot be regarded as a reliable guide to the situation obtaining in medieval times. For one thing, considerable tracts may be presumed to have disappeared as a result of centuries of

piecemeal enclosure. And even those field systems which survived
relatively intact had often changed radically through time, alike in their
usage, layout and nomenclature. The clearest illustration of this
transmutability is provided by the field system of Church Bickenhill,
which can be shown to have passed through at least three identifiable
post-1300 phases (Figure 21).

In high medieval times (Figure 21a), despite the fact that the manor
covered only 738 acres (299 hectares), we are confronted with a
bewildering complexity of many small open fields and furlongs. To
the north of the village, in addition to one main field, Trowemorfeld —
which included Longeforlong, Mereforlong and Brehcforlonges — there
were at least three apparently independent open-field units. In 1376-7
Heystonweforlonges is described as 'a common field of cherche
bekenhulle . . . called ye heystowe'. Similarly, in the fourteenth century
Pipershale forlonges has the appearance of being a separate field,
containing indeed its own subsidiary furlongs: 'le longeforlong in
Pipereshale' and 'that furlong which is called le Scherccbuttes . . . in
piperhushale'. Lastly, Westanefordemor furlong, though ultimately
absorbed into Trowemorfeld, is in the fourteenth century never
connected with it, but features rather as an independent unit.

To the south of the village, the second of Church Bickenhill's three
main fields was Le Leefeld which at this time included Le Redelondes,
Le Burysonde, Le Stockynges, as well as Waterschypeforlonges. Holforde
(furlong) may also have been part of Leefeld, but Le Hongingehull was
almost certainly a separate entity. The third main field, Haregrove or
Haregrofylde, lay to the west. Lutle Feld, despite its importance later,
is only twice referred to in fourteenth-century deeds. On one occasion
it appears to stand on its own, but on the other it is enigmatically
referred to as the 'little field [*parvo campo*] which is called Haregreve'.

Apart from Haregrovemor, the manor of Church Bickenhill itself was
seemingly without common pasture, but by the early seventeenth
century, and probably long before, its inhabitants had the right to graze
their cattle on Hill Bickenhill's Bicknell Heathe. The two common
meadows that can be placed, Pipershalemedewe and La Le Medwe,
were at the end of their respective open fields, where these sloped down
towards brooks; but the unplaced Dowse medowe and Brodmedowe are
both described as being in Haregrofylde, and were probably on either
side of the alluvium-banded stream that runs through it from south to
north. (Skipp and Hastings, 1963, 15-20)

There is no way of discovering whether — and still less, in what way
— the six or seven apparently separate open fields were combined for

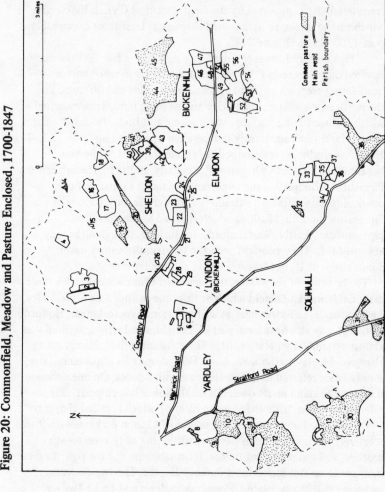

Figure 20: Commonfield, Meadow and Pasture Enclosed, 1700-1847

Key to Figure 20

Yardley

1. Stichford Field 2. Fell Meadow 3. Church Field 4. Wood Meadow 5. Stockstile Field 6. Crabtree Field 7. Acocks Green Field 8. Showell Green 9. Wake Green 10. Moseley Wake Green 11. Swanshurst Common 12. Billesley Common 13. Yardley Wood

Sheldon

14. Outmoor Green 15. Kitts Green 16. Rye-eddish Field 17. Elder Field 18. Ridding Field 19. Garret's Green 20. Radley Moor 21. Wells Green 22. Great Sheldon Field 23. Greatock Field 24. Great Hatchford Field 25. Little Hatchford Field

Lyndon

26. Round Meadow 27. Middle Meadow 28. Line Field 29. Line Meadow

Solihull

30. Solihull Wood 31. Shirley Heath 32. Lode Heath 33. Seed Furlong 34. Hain Field 35. Berry Field 36. Wheat Croft 37. Common Meadow 38. Copt Heath

Bickenhill

39. Middle Field 40. Little Field 41. Marston Culy Green 42. Broad Meadow 43. Far Field 44. Wavers Marston Common 45. Bickenhill Heath 46. Near Troughmoor Field 47. Far Troughmoor Field 48. Hawthorn Meadow 49. Hawthorn Field 50. Lake Meadow 51. Well End Field 52. Little Field 53. Archall End Field 54. Watershutt Meadow 55. Bakehouse Meadow 56. Hanging Hill Field

cropping purposes in the fourteenth century. By the early seventeenth century, on the other hand (Figure 21b), the various field areas had been brought into a single integrated system: so that what we find in effect, as well indeed as in name, is a fully rationalised three-field system embracing an area of not less than 600 acres (243 hectares). A deed of 1612 specifically describes 30 acres of arable in Church Bickenhill as being 'in the Three Common feildes there', giving their names as Hargrave feilde, Watershipp feild (cf. Waterschypeforlonges) and Troughmoore feilde. Moreover, although in this deed no less than twelve component furlongs are particularised, by 1677 when an extremely detailed survey was made, all furlong names seem to have been discarded. Other changes which occurred some time between the fourteenth and the seventeenth centuries included the introduction of hades, which were used for the grazing of tethered animals and lay dispersed among the arable selions, and the making of several new common meadows.

Meanwhile, something like a third of the commonfield land had been enclosed: a process which continued during the seventeenth and eighteenth centuries until by the parliamentary enclosure of 1824 virtually the whole of the centre of this once extensive field system had been withdrawn into severalty (Figure 21c). And it was this process of

Figure 21: The Church Bickenhill Field System, 1300-1824

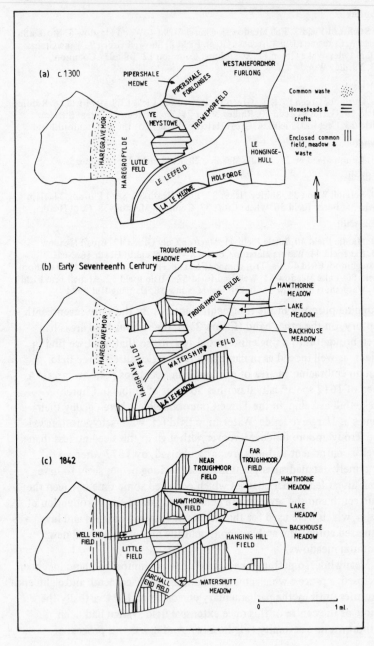

piecemeal enclosure which in turn led to most of the changes in
nomenclature that occurred between 1677 and 1824. Thus, a fragment
of north Hargrave was by 1824 known as Well End Field; virtually all
that survived of Watershipp as Archall End Field; and even the better
preserved Troughmoor had apparently come to be thought of as three
separate entities and bore three different names. In other words, having
been rationalised into a 'classic' three-field system in the early modern
period, before it finally came to an end, the Church Bickenhill system
had relapsed back into an appearance of complexity not dissimilar from
that which was found in medieval times (Skipp and Hastings, 1963,
20-9).

Church Bickenhill is a particularly well documented manor. And it
cannot be pretended that all the *c*. 1300 field systems which have been
conjecturally reconstructed (Figure 22) by the correlation of early
documentation with nineteenth-century enclosure awards and tithe
surveys are put forward with an equal degree of confidence. With the
manor of Solihull (*née* Ulverley), for example, the information available
is very slight indeed. At Elmdon, where enclosure was complete before
1650, none of the medieval furlong names come through on to the
eighteenth- and nineteenth-century maps, so that — apart from the
support of late surviving ridge and furrow — the location of the
'original furlongs' is entirely inferential. Again, in reconstructing a good
number of the field systems, for the sake of achieving a reasonably
complete impression, quite a lot of post-medieval documentary
information (indexed in brackets) has had to be interpolated; and on
occasions even greater risks have been taken. At Lee, to cite but one
example, there are no references to Bloomfields in medieval times; nor is
there proof that it ever in fact served as commonfield. On the other
hand, it is shown as a strip field with dispersed ownership on an estate
map of 1756,[2] and is therefore entered on Figure 22 with a
questionmark.

No less unfortunate is the fact that several suspected or known
commonfield systems cannot be reconstructed at all. That Ulverley
originally possessed one can be inferred from remarks made by Dugdale
(Dugdale, 1656), while its location is suggested by the presence of four
large adjacent fields known as Wolvey Fields in the 1839 Bickenhill
tithe survey (Skipp, 1977, 1978). But specific medieval references are
lacking, so no details can be recovered. At Elmdon one is confronted
with the opposite difficulty: apart from the four mapped furlongs, the
names of several other commonfields are known, but there is no means
of establishing their position. Yet another regrettable circumstance is

Figure 22: Conjectural Open Field, Common Meadow and Common Pasture, c. 1300

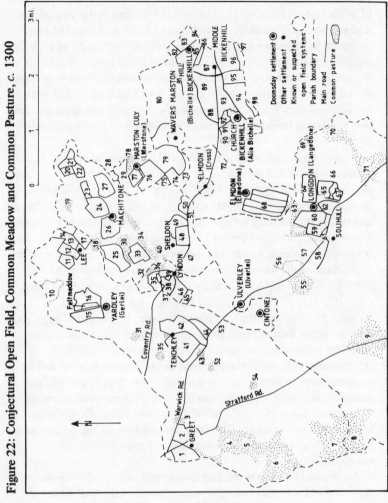

Key to Figure 22

1. (Heanfield) 2. Gravel Fields 3. (Maule Green) 4. (Greet Common) 5. (Waste called Swaneshursteland) 6. (Billesley Common) 7. Yardley Woods 8. Solihull Woods 9. (Shirley Heath) 10. Rudyng 11. Wodemedwe 12. Honehull 13. Atmunchull 14. (Bloomfields?) 15. Stycheforde Felde 16. le Churchefeld 17. Longefield 18. Kitt Greene 19. Le Outmor 20. Asheforlonge 21. Holifast 22. Monland 23. (Berriatts) 24. Ruyedissh 25. Cokeshote Feld 26. Elrefurlang 27. Rudyng 28. (Stockinges Meadow) 29. Le Rudyng 30. Garardes Grene 31. (ffast Greene) 32. (Lindon Green) 33. Asshewell 34. (Radley Moor) 35. (Chappell ffield) 36. (Lady Wood) 37. (Round Meadow) 38. Mydle Fyelde 39. (Poole Field) 40. (Outwoods) 41. (Over Hayinfield) 42. (Nether Hynefield) 43. (Flint Green) 44. Wyssysclowes medewe 45. (Lyne meadow) 46. Lyndon ffield 47. (Wells Green) 48. (Sheldon ffield) 49. Greteokesfeld 50. (Great Hatsford) 51. (Little Hatsford) 52. (Fox Green) 53. (Acocks Green) 54. (Hawee Grene) 55. le Graungeheth 56. Crabewallefeld 57. Orchardesfeld 58. Le Chirchefeld 59. Longeforlong 60. Bromfild 61. Teynteres Grene 62. Hem Furlong 63. Heath of Elmedon 64. Sydefurlong 65. Le Burifeld 66. (Wheatefield Meadow) 67. Watcroft 68. Original furlongs? (Elrefurlang, Auelangfurlang, Ethfurlang, Foxoles) 69. (Katherine de Barnes Heathe) 70. Hinewadsheth 71. Cophethe 72. Heath of Dunstal 73. (Elmdon Meadow) 74. (Light Wood) 75. (le Brodemede) 76. (Middle Field) 77. Luttelfeld 78. (Marston Culy Green) 79. (Far Field) 80. (Bicknell Heathe) 81. (Halywellmore) 82. Corsedeparoc 83. le Paroc, le Smalihet 84. Wyllemedewe 85. Helywellefeld 86. le Brodemeduwe 87. Echelesfeld 88. Pipershalemedewe, Pipershale forlonges 89. Westanefordemor furlong 90. Hargrovemor 91. Haregrofylde 92. le heystowe 93. Trowemorfeld 94. La Leefeld 95. Le Hongingehull 96. Hasilfeld or Soutfeld 97. Blacalremeduwe 98. La Le Medwe *Names in Brackets derive from post-1550 sources.*

the fact that most of the common wastes, although sometimes supplied with medieval names, because of the dearth of boundary descriptions, can only be represented in terms of the areas shown on eighteenth- and nineteenth-century maps.

The most that can be said for the *c.* 1300 map, therefore, is that despite considerable doubts over many points of detail, it ought to be sufficient to provide some indication of the extent and layout of communally worked land units at this time; and that, hopefully, it may be considered adequate to sustain at any rate the general direction of the argument that follows (Skipp, 1960, 6-24; 1970, 5-18, 21-35; 1977, 3-9, 15-16; Skipp and Hastings, 1963, 8-33).

By 1300 there were probably 16 or 17 individual sets of open fields in the 26,284 acres (10,637 hectares) covered by the five parishes; and with their associated common meadows and wastes, there can be little doubt that they were being operated on a communal basis. From 1294 onwards many land charters actually use the term 'commonfield'; several speak of dole meadows, like the Yardley deed of 1290 which grants 'two selions of land with one share of meadow', or that of 1406 concerning 'seven selions of land in le Churchefeld and one dole of meadow in le Feldmedewe'. Elsewhere common grazing seems to be implied, as in the late thirteenth-century Hill Bickenhill grant which includes 'free pasture in the heath towards le eccsle and in edymeccles and in pasture of Westoneford and in pasture of Hellywellmore'. Holdings are invariably dispersed, like that described in 1345 as '16 selions lying scattered in the common field of Yerdeley'; and the not infrequent use of the term 'rood' as an alternative for selion suggests a unit approximating to a quarter of an acre. This would tie in with the fact that surviving ridges in these and neighbouring parishes have a tendency to be about 200 to 220 yards long and 5 to 6 yards wide.

It is notable, however, that individual field systems were extremely small, most of them falling between 100 and 200 acres (40 and 81 hectares). They are also, without exception, located either in the Tame-Cole-Blythe Basin, or else on the very edge of the Arden Plateau.

Yet another characteristic is the fact that each individual open-field system was associated with a nucleated or semi-nucleated settlement. Indeed, despite their diminutive size, at Bickenhill all five open-field settlements were in origin, or became, independent manors. By contrast, the large tracts of Solihull and Yardley where open fields are not to be found are also devoid of nucleated villages or hamlets, being characterised rather by twelfth- and thirteenth-century farms held in severalty, with their isolated and often moated farmsteads (Skipp, 1970, 24-35; 1977, 15-16). This sharp dichotomy suggests that in these Arden parishes, whenever and however open-field agriculture evolved, it did so exclusively in pieces of country which were among the earliest to be cleared and cultivated.

Open-field agriculture, alongside its connection with nucleated settlement, would also appear to have been anciently associated with villeinage. For, although the earliest available documentation shows that by 1300 an appreciable number of freeholders were involved with open-field land, at Bickenhill and Sheldon, where this land predominated, the majority of peasants ultimately held by copy: whereas in early fourteenth-century Yardley free tenants probably outnumbered customary by 2:1, and at Solihull in 1632 there were only 5 copyhold and 9 leasehold tenures, as against 75 free tenancies (Skipp, 1960, 27; 1970, 36-7; 1978, 7). Since it is known that the makers of the twelfth- and thirteenth-century isolated farms were invariably freeholders, and since in any case only one possible freeman is to be found in the Domesday entries for the five parishes, this link with villeinage may be taken as a further pointer to the antiquity of the open fields.

For the purpose of looking at the topographical development of the north Arden open-field systems prior to 1300, it would seem sensible to concentrate at first on the nine settlements which, on place-name and other evidence, can be identified with Domesday entries (Skipp, 1960, 9; 1970, 1). These have been indicated on Figure 22 by the use of a distinctive symbol. As will be seen by comparing Figures 19 and 22, with only one partial exception – Church Bickenhill – the open fields immediately adjacent to Domesday settlements lie on restricted areas either of sand and gravel or of Boulder Clay/undifferentiated drift.

Four settlements – Ulverlei, Cintone, Elmedone and Langedone – found such locations on the spurs extending out from the Arden Plateau to which reference has already been made. At all four of these sites an ancient road spines the top of the ridge. This probably originated as the headland or meer between the original furlongs, which – where they can be mapped – appear to have been laid out laterally along the length of the ridge, with their selions running down to an alluvial meadow and stream on either side. The name Langedone actually means 'long hill'; and the Elmedone and Ulverlei sites could equally well have been so described.

Bichehelle (Hill Bickenhill) used a similar spur at the south-eastern extremity of Bicknell Heathe. Again, a ridge road divides the two main field areas: the antiquity of which is confirmed in this case by a reference to it in 1304-5 as 'a way called le holewewey'. In this case also both of the common meadows which ran along the flanking watercourses can be named and plotted. Merstone exploited an isolated sand and gravel patch, but in other respects its situation was remarkably similar. So was Yardley's, though here the sand and gravel lay in a shallow depression, rather than on a ridge. The fields therefore occupy rising ground on either side of a small stream with a common meadow and the usual trackway separating them. Machitone was slightly different again. Although based on an interfluvial ridge, this was much broader than its counterparts elsewhere, while the vital veneer of light drift soil was to be found on only its southern side. Nevertheless, here also the main sand patch seems to have been divided in such a way that it could accommodate the two presumed original fields (Elrefurlang and Ruyedissh).

Looking now at the open fields attached to settlements which cannot be identified in the Domesday survey, it can be seen that in the two large parishes that lay athwart the Arden Plateau these show a predilection for the same kind of situation. Solihull's open fields lie like a saddle across the same sand-capped ridge as was used by Longdon. In fact, they abutt the latter's fields immediately to the west. In the manor of Yardley, Tenchley and Greet also used ridges near the northern extremities of the Arden Plateau: the former on Boulder Clay, the latter on sand and gravel. Greet is derived from OE *grēot* meaning 'gravel'. Even the hamlet of Lee, which used valley land, was established on a small patch of Boulder Clay half a mile east of the village of Yardley and immediately overlooking the Cole.

But if Solihull and Yardley could relate all their open-field settlements to the lighter soils, the smaller parishes, which were also the ones that were confined to the lower ground, clearly could not. Wavers Marston, whose field system cannot be reconstructed, may have exploited the western edge of the otherwise barren and hungry Bicknell Heathe; and Lyndon's open fields were partly laid out on Boulder Clay. But like the founders of the Domesday alia Bichehelle (Church Bickenhill), the settlers at Middle Bickenhill, Elmdon (Cross) and Sheldon had no alternative but to tackle the clearing of the intractable Keuper Marl.

Turning now from the location and siting of the open fields to their form, broadly speaking, three types of layout can be distinguished:

1. Compact or relatively compact groups of two, three or four small fields, as at Longdon, Hill Bickenhill, Middle Bickenhill, Marston Culy (Merstone), Lyndon, Sheldon, Yardley, Tenchley and Greet.
2. Extremely compact groups of five or more fields, as at Church Bickenhill and Solihull.
3. A group of three small core fields, with several outlying fields that are separated from the core, as at Machitone.

So far as the evolution of these layouts is concerned, in the majority of cases it seem likely that the original nucleus would have consisted of two units, which, where possible, as we have seen, were made by the bisecting of a small area of glacial drift. And indeed, with several settlements open-field development appears not to have progressed beyond this initial stage: for instance, at Tenchley and Greet, where there was a seeming reluctance to extend this form of cultivation beyond the bounds of the drift, and at Middle Bickenhill, where the scope for expansion was lacking. Incidentally, Roberts found a similarly inhibited nucleus at Tanworth in Arden, which utilised a gravel spur on the southern rim of the Arden Plateau, and apparently only ever possessed two small open fields (Roberts, 1968, 103-4, 111; 1973, 226).

Settlements which did go on to extend their open-field arable further may next have added a third field. This can probably be seen at Yardley and Machitone where in both cases the new field was partly located on a small nearby area of drift but also involved clearing a certain amount of marl. In view of this fact, it may be more than coincidence that at both places this third field was called Rudyng, from OE *hryding,* 'clearing'. And if, as seems likely, the two original fields at Marston Culy were Luttelfeld and the northern furlongs of what ultimately became Middle Field: then, with another conveniently placed open field called Le Rudyng, a virtually identical progression could be posited there.

The third and final notional phase would then be the addition of still further fields or furlongs,[3] either immediately adjoining those already existing, as at Church Bickenhill and Solihull, or separated from them, as at Machitone.

However, at any point in this progression, it would seem that another course of action was also possible: namely, the establishment of a completely new 'colony' settlement, with its own entirely separate group of open fields. This might be brought about, amongst other things, by shortage of space, the formidableness of the immediately surrounding marl-based forest, or because the distance between the old settlement and the new land which was to be cleared would have been excessive.

We would almost certainly be correct in inferring that Lee, Tenchley and Greet originated as secondary settlements which were founded in this way from the parent village of Yardley. And in all three instances it looks as though the principal motive was to leap over the Keuper Marl. At Sheldon, by contrast, which can be presumed to have originated as a colony of Machitone, the main factor would have been locational. In this predominantly marl-covered parish, the forest simply had to be reclaimed. But it would have made no sense to tackle the distant southern clays from the homesteads of Machitone.

In reviewing the various factors which may have contributed to the development of the open fields, Baker and Butlin concluded that the two main processes involved were probably the subdivision of once individually owned closes due to partible inheritance or the need to accommodate expanding numbers, and communal clearance followed by co-aration (Baker and Butlin, 1973, 635-41). Locally, partible inheritance is unlikely to have been a major factor. By the early seventeenth century certainly, primogeniture or in the case of Longdon ultimogeniture (Borough English) was the rule so far as customary holdings were concerned: and the former appears to have been the usual

practice among freeholders (Skipp, 1970, 107-9). All the same, two or
three cases of the shared ownership of enclosed land have been noted at
Yardley. Thus in 1345 John atte Forde was granted 'four plecks of
land . . . lying in a field called Solerfelt'; and seven years later John de
Yardley received 'two selions of land and a gore in the Masones croft'.
Almost certainly these two fields originated as individually owned
assarts of the Soler and Mason family respectively (Skipp, 1970, 43).
But although they may have become strip fields, there is no evidence to
suggest that they evolved further into full communal ownership.

Roberts, it is true, did find such evidence at nearby Coleshill, where
a close which was granted to Simon March in the mid-thirteenth century
had by 1300 become a strip field in the possession of eleven different
tenants, and was ultimately incorporated as part of the Coleshill
commonfields. But Roberts himself regarded this example as
exceptional, 'and not necessarily applicable to the larger areas of open
field which may have been the product of communal activity' (Roberts,
1973, 638). So far as the bulk of open field in Arden is concerned,
therefore, although an evolution from communal assarting and co-
aration could never be proved, so much circumstantial evidence seems
to point in this direction that at the present time it would at least
appear to be the most likely available explanation.

Secure knowledge of the chronological development of the local
field systems is almost equally lacking. Nevertheless, certain general
points can perhaps be made. To begin with, the dearth of evidence for
Roman or prehistoric settlement makes it difficult to propose an origin
for the open fields — or even for the forest clearings on which they
were based — that reaches back before the Saxon settlement. At the
same time, however, the drift-capped spurs and isolated sand and
gravel patches with which the majority of open fields were associated
may be seen as the natural — almost, one might say, the inevitable —
primary settlement points in this terrain. And therefore, like the river
gravels of other areas, they are precisely the locations where, on *a priori*
grounds, one would expect any early agrarian development there might
have been to have taken place.

That many of the open fields, or at any rate their clearings, were in
existence by the end of the Old English period is suggested by the
already noted one-to-one correlation of all Domesday settlements with
such areas. And the possibility of a Saxon connection gains further
support from the fact that Bickenhill, the parish which was far and

away the most densely settled by 1086, was also the parish which can be shown subsequently to have possessed by far the highest proportion of open-field land.

The Saxon exploitation of the upper Tame-Cole-Blythe Basin may not have got under way until the seventh or eighth century, and the earliest documentary reference to a five-parish settlement does not occur until 972, when Yardley is recorded in the Pershore Charter as having five manses, and mention is also made of Cintone in the accompanying boundary survey (Skipp, 1970, 9). The five manses presumably imply that the *Gyrdleah*, or 'yard clearing', where the open fields were later to be found was already under cultivation. So if the fields were in some sense the creation of the original settlers, they would belong to the mid Anglo-Saxon period.

Arguably, there is a further factor which tends to increase the likelihood that the open fields derive from the same point in time as the settlements with which they are connected. This is the small-scale — and indeed, short-term — nature of the thinking which seems often to have been involved in their initial siting and layout. Not only were most of the drift areas that had to be utilised extremely tiny, it would seem doubtful whether the first cultivators always contemplated using even the full extent of these as arable land. And certainly, at Yardley, Machitone and Hill Bickenhill, it is noticeable that the drift areas, although bisected for cultivation purposes, were not in fact divided into two equal parts, as one would have expected if there had been an intention from the start to take in the whole of the area in question. Perhaps, too, the remarkably close siting of the Anglo-Saxon settlements of Ulverlei and Cintone may be regarded as another indication that, along with the settlements of this once heavily-wooded and therefore uninviting piece of country, the majority of its field systems must have had exceptionally tentative and unambitious beginnings.

Apart from the nine field systems that can be linked with Domesday settlements, it seems probable that at least one other had its origin in pre-Conquest times. The actual town of Solihull most likely began as an *ab ovo* market borough about 1170, being founded from Ulverlei which thereafter became known as Olton, or the 'old town' (Skipp, 1977, 10-11). All the same, its open fields, which were situated immediately north of the new town, most probably came into existence at a much earlier date. For although Ulverlei had only 8 plough teams in 1086, it had 20 ploughlands, while the value of the manor in 1066 had been £10, as against £4 at the time of the survey. Clearly, in late Saxon times this manor had had a particularly large area of arable land under cultivation, and – in view of their characteristically early location – it is difficult to believe that what later became Solihull's open fields would not have formed part of it. Still more tellingly, the presence of the reversed-S aratral curve in the boundaries of some of Solihull's burgage plots implies that, as at Stratford-upon-Avon, the new town was itself laid out on pre-existing open-field land.

So probably at least ten of the sixteen or seventeen known open-field systems may be regarded as in some sense having been rooted in the Saxon period. However, if any credence may be placed on the Domesday statistics for Sheldon and Yardley, it seems no less likely that the process of open field making must have continued into post-Conquest times. When the figures for its three Domesday vills are added together, Bickenhill, which, excluding its later accretion Lyndon, covered 2,838 acres (1,149 hectares), is credited with 11 ploughlands, 9½ plough-teams, 19 villeins, 6 bordars, and was valued at £4. Machitone (Sheldon), with 2,500 acres (1,012 hectares), had only 5 ploughlands, 3 plough-teams, 10 villeins, 4 bordars, and was valued at £1. The Yardley figures are unfortunately amalgamated with those of Beoley, but the two manors together could only muster 8 villeins, 10 bordars and 1 radman with 9 ploughs. And Yardley by itself was over twice the size of Bickenhill. These figures hardly seem impressive enough to justify the assumption that the communal settlement of Sheldon, as well as Machitone, had been founded by 1086; still less that the Domesday manor of Yardley, in addition to the parent settlement, included the open-field hamlets of Lee, Tenchley and Greet. Perhaps it is also worth noting that, whereas the extreme compactness of the Bickenhill and Solihull open-field systems would not argue against the idea of a fairly rapid process of evolution, at Machitone the block of common pasture (Kitt Greene and Garardes Grene) that separates Elrefurlang from the outlying Cokeshote Feld and Asshewell points rather to a lengthy, and indeed interrupted, agrarian expansion. For, since this waste cannot be explained geologically, it could well signify a pause in arable development that was sufficiently protracted for common pasture rights to have become irreversibly established over the land in question.

When then did the process of open field making come to an end? Roberts argues that at Tanworth all clearing after 1220 took the form of enclosed severalties (Roberts, 1973, 228). The making of private assarts must have been well advanced in our own parishes by then. For in 1221 what can only be described as a collective rearguard action against this practice led to four Yardley enclosure disputes being taken before the Justices in Eyre at Worcester in a single year (Skipp, 1970, 26-8). In one case Thomas Swaneshurst had made an assart, but 19 of his neighbours had pulled down his hedges. Undoubtedly the underlying reason for this violent behaviour was that so much clearance had taken place by 1221 that — at least in the opinion of the offending peasants — there was now a serious threat to their supply of pasture. Against this background, it seems likely that the communal hamlets of Lee, Tenchley and Greet would have been twelfth- rather than thirteenth-century foundations. And certainly the subsidy roll of 1275 proves that they had become well-established and relatively populous settlements by that date (Skipp, 1970, 25).

In general, the conclusions of this paper tie in closely with those advanced by Roberts in his wider survey. However, Roberts, from his more distant vantage point, drew a distinction, based on the 400-foot contour, between what he termed the 'woodland fringe systems' of the Trent-Tame-Blythe sub-region and the 'woodland core systems' of the uplands (Roberts, 1973, 209-10). Such a distinction hardly seems useful in the area covered by the five parishes. For the field systems that lie just above the 400-foot mark do not differ in any sharply definable way from many of those that lie below it.

It is true that if one thinks in terms of *parishes*, Bickenhill, Sheldon and Elmdon, which belong exclusively to the lowland, had a much higher proportion of open-field land than the two parishes extending up on to the Arden Plateau. But as we have seen, it was not the parish but the communally settled village or hamlet that generated the open fields. So, rather than allocating the parish of Sheldon's system(s) — as Roberts does — to the woodland fringe, and the Longdon system to the woodland core (Roberts, 1973, 220), it would probably be less confusing to categorise the three nucleated settlements of Machitone, Sheldon and Longdon as all having had a woodland fringe system.

And indeed, if one thinks of the nucleated settlements perched on the very edge of the Arden Plateau as belonging to the woodland fringe, then in this western part of Arden at least, one would be left with no woodland core open-field systems at all. For, apart from its periphery, the whole of the Arden Plateau seems to be devoid of ancient nucleated settlements and their accompanying open fields. As Figure 18 shows, the bulk of this higher land was shared between four large parishes: King's Norton, Yardley, Solihull and Tanworth. Paradoxically, their size is an index of their lack of early development. In Domesday times huge areas of woodland may be attributed to them: excluding

Kings's Norton, perhaps as much as 12 leagues long by 5 leagues wide.[4] This was the true post-Conquest woodland core. And communal assarting – for reasons which Roberts himself has discussed (Roberts, 1973, 229) – had passed out of vogue by the time it came to be cleared. Thus, historically speaking, the line which defines the extremity of open-field land round the Arden Plateau is of greater significance than the 400-foot contour. For it represents, probably in fairly precise form, the extent of Saxon and Norman settlement and agrarian development in this part of the English midlands.

Not that the plateau top itself was without a part to play in the region's early economy. The boundary between the dioceses of Worcester and Lichfield ran across it; and this in turn is probably to be equated with the boundary between the Hwicce and the Mercians; and before that, Ford has hinted, even between the Iron Age Dobunni and Cornovii – hence perhaps the siting of Berry Mound (Ford, 1976, 277-9). Again according to Ford, from relatively early in the Saxon period, settlements from as far away as the Warwickshire Feldon may have been intercommoning to the south of this ancient boundary, and those from the Tame-Cole-Blythe Basin to its north (Ford, 1976, 279-82). Perhaps it was the grazing of Saxon, if not of Iron Age, cattle which in places led to the early degeneration of the lighter oak-birch woodland into heath. But by 1200 the manorial lords who had come to control the Arden Plateau had also established their right to develop it. And within a century or so, under their aegis, a new and absolutely distinctive kind of agrarian landscape had been brought into being.

This clarifying of the distinction between the 'woodland core' and 'woodland fringe', so far as the Arden Plateau and Tame-Cole-Blythe Basin are concerned, could be of value in the present context. For, to borrow from scientific terminology, it may enable us in a stricter sense than might otherwise be the case to regard the former as a 'control' on the latter: and in so doing, strengthen some of the findings and perhaps widen the usefulness of this paper.

Notes

1.   Geological Survey of Great Britain, Sheet 168 (Birmingham) Drift.
2.   Map of Estates in . . . Sheldon and Coleshill . . . belonging to John Taylor, 1756, Birmingham Reference Library, 435851.
3.   The two terms are sometimes used indiscriminately in the early documents, and even interchangeably. Generally the term 'field' seems to imply a certain independence of status, whereas 'furlong' could either be an independent field

unit, or merely a component part of such an entity.

4.   Yardley (with Beoley) had woodland 6 leagues long, 3 leagues wide; Longdon 1 league long, ½ league wide; Ulverley 4 leagues long, ½ league wide. Tanworth is thought to have been a member of the Feldon manor of Brailes and probably ought to be credited with most of its woodland 3 leagues long and 2 leagues wide.

9 THE ORIGIN OF PLANNED FIELD SYSTEMS IN HOLDERNESS, YORKSHIRE

Mary Harvey

In 1969, Matzat and Harris drew attention to an unusual pre-enclosure field pattern which was found in the extreme eastern part of the former county of Yorkshire (now Humberside) (Matzat and Harris, 1971). So different was this layout from the one usually associated with open-field farming in lowland England that it raises important questions concerning origins and development. Is the form an adaptation to a distinctive physical environment, or did it result from particular social and economic conditions prevailing in the area in the past? This paper speculates upon some of these problems and considers possible explanations.

A Planned Field System

Holderness is a low-lying region to the east of the Yorkshire Wolds, whose characteristically hummocky topography is derived largely from Pleistocene drift deposits. Much of the area was ill-drained until the eighteenth and nineteenth centuries, and extensive areas of marshland occurred in the west, in the region of the River Hull valley, and in the south, associated with the River Humber and with four small streams draining into it. Settlements and arable land were nearly always located on the higher ground, whereas meadow land lay towards the margins of a township. Except in the west and south, however, meadow was of small extent, and arable fields occupied by far the greater part of a township's area.

Enclosure of the arable land occurred relatively late in Holderness, some 55 per cent of townships still lying open by the early eighteenth century. The structure of the fields is thus fairly well known, and it is clear that by the seventeenth and eighteenth centuries this was quite different from the complex patchwork of furlongs and strips associated with the Midland system. Three features in particular are worthy of note:

1. Townships usually had only two arable fields, despite the large area

184

devoted to arable cultivation.

2. Many strips within the fields were of great length, often extending from one field boundary to another, a distance of over one mile in some townships.

3. Strips lay parallel throughout the greater part of a field, and as a result frequent changes in the orientation of strips were absent. Furlongs, where these occurred, were few in number within any one field, and were usually functional rather than physical divisions within the fields.

There are few maps of Holderness open fields, but the five which are known all exhibit these characteristics in some degree. A map of an estate in Skeffling, in south-east Holderness, made in 1721, names three fields, West Field, In East Field and Out East Field, but in every other respect the layout resembles the Holderness 'type' (Figure 23).[1] All strips lay parallel, save for a block of strips to the north of the village, and a majority extended for over half a mile from one field boundary to another. The absence of furlongs is very striking. On the earliest map, a 1636 plan of Winestead in south Holderness, two fields are shown, East Field and West Field.[2] Although only part of the internal structure of these fields is detailed, it can be assumed that the majority of strips had a common east-west alignment.

For the remaining Holderness townships, evidence of field structure must be derived from written documentation. No comprehensive survey of the open fields of any village has yet been found, but in a number of cases terriers are sufficiently numerous to enable fairly detailed reconstructions to be made. The arable land of Preston thus lay in two large fields, North Field and South Field, whilst meadow land occupied the low-lying west and south-west of the township (Figure 24) (Harvey, 1978). Except for a small area in the extreme east of South Field, all strips were orientated north-south, at right angles to the village enclosures, and many reached from one field boundary to another. Only seven subdivisions occurred in each field, and these were not marked by any obvious change in the alignment of strips.[3]

Similar simplicity of form typified the fields of almost all other Holderness townships whose seventeenth- and eighteenth-century layouts can be reconstructed. It seems clear, moreover, that this was not of recent origin. Documentation becomes less plentiful and more general before the seventeenth century, but there is nothing in this earlier material to indicate that different arrangements prevailed. Townships with two arable fields were again common in the medieval

Figure 23: Mr Bee's Estate in Skeffling in 1721

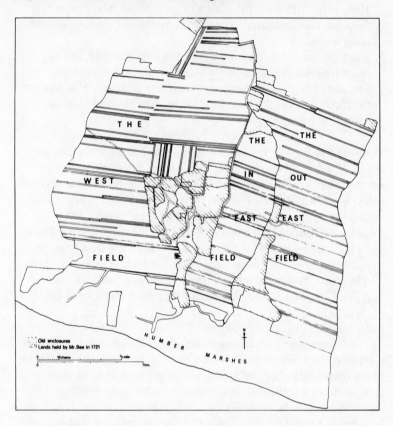

Source: Based upon a map in the Humberside County Record Office DDCC(2)
G (2).

Figure 24: The Field Structure of Preston in 1750, Showing the Approximate Location of the 'Chantries' Land

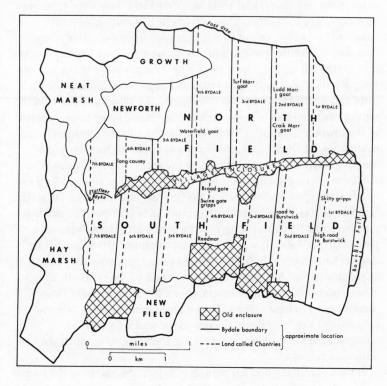

period, and references to strips invariably show that these could be of great length. In Long Riston, for instance, there are several examples of strips in the two fields, East Field and West Field, extending from one field boundary to another. Thus, a thirteenth-century land grant described two strips, one of which 'extended from the village to the boundary of Routh' (the township to the west of Long Riston), and the other extended 'to the boundary of Rise' (the township to the east of Long Riston).[4] Similarly, East Halsham had two arable fields, North Field and South Field: the former adjoined the village enclosures in the south and a small area of low-lying meadow ground in the extreme north of the township, the latter extended from the village enclosures to the boundary with Ottringham and Winestead in the south.[5] Frequent mention was made of strips extending the entire length of these fields.

The medieval evidence indicates, therefore, that in all townships for which meaningful documentation has survived, the extent of the arable land was as great in the thirteenth century as in later periods, and that the distinctive field morphology of the period after 1600 was also present before 1600. It would be unwise to assume from this that no changes had occurred in field layout over this time, especially since it is impossible to find one township where details of the fields are known in both post-medieval and medieval times. Nevertheless, very close parallels seem to have existed. The evidence would further suggest that the origin of the form lies in the period for which no detailed documentation is available, and that it is therefore earlier than the mid-thirteenth century.[6]

Before examining possible periods of origin, the nature of the form itself must be considered. Its simplicity and uniformity suggests a planned origin, rather than evolutionary development. The marked absence of evidence for the slow accretion of new land to an established core, the lack of furlongs and of formal complexity around the margins of the fields all belie a system which has grown slowly from small beginnings. Yet all these aspects would be in no way unusual if the fields had been planned and laid out at one point in time. Could it be that in Holderness we have an example of field planning on a scale which is apparently unknown elsewhere in Britain, but for which there are a number of examples on the continent (Filipp, 1972; Nitz, 1971)? Significantly, many of these continental layouts closely resemble Holderness forms.

If Holderness fields are indeed a result of deliberate planning, then several important questions must be asked concerning the background to such planning. When, and by whom was planning undertaken? What was the reason for it? How was field planning effected and what did the new layouts replace? In the absence of any documentation the answers to all these questions must be a matter for speculation, but the distinctiveness of the form in Britain demands that they be given some consideration.

Origins and Causes of Planning

Studies of both field and village planning in continental Europe (Göransson, 1958 and 1969; Filipp, 1972) and of village planning in northern England (Roberts, 1972; Sheppard, 1974) have highlighted the all important role of a landowner, or some superior authority, in initiating and controlling change. It seems reasonable to suppose that this might also have been the case in Holderness. Furthermore, in order to explain the standardisation of layout throughout the region it may be necessary to identify a period not only when lordship control was strong, but also when the whole area was held by one, or very few landowners. A complex tenurial structure would not only make the implementation of such a uniform field layout more difficult to achieve, but also much more difficult to explain.

Although details of field morphology cannot be traced beyond the thirteenth century, information regarding landholding has survived for a longer period, albeit in discontinuous form.[7] This evidence suggests that there was only one period when the conditions thought to have been necessary for planning were present, namely the decades following the Norman Conquest of 1066.

Norman rule had resulted in the eradication of the old order of landownership in England, Norman lords replacing Anglo-Saxons almost without exception (Stenton, 1971, 625-6). In the case of Holderness, Domesday Book shows that the rather complex pattern of small estates and multiple lordship prevailing in the mid-eleventh century was replaced by a much simpler situation. The whole area, apart from lands held by the Archbishop of York and the canons of St John of Beverley, became a single honour or lordship, held initially by Drogo de Bevrere, and subsequently by the earls of Albemarle (Poulson, 1840). This unity could not have lasted long, however, for grants of land to religious houses with subinfeudations by successive overlords quickly resulted in a more complex landownership structure. If, therefore, tenurial unity was a prerequisite for planning, the likelihood is that opportunities for its implementation diminished rapidly through the twelfth century.

On the other hand, it is doubtful whether unity of lordship alone would account for the dramatic changes envisaged. More likely, there were also specific reasons why purposeful action by the lord was either necessary or desirable. In eleventh-century Holderness, these may well have concerned both the economic and the tenurial state of the region.

Domesday Book statistics suggest that in 1066 Holderness was a relatively prosperous and developed area. Despite large areas of ill-drained, low-lying land, it was well settled and probably also moderately intensively exploited (Sheppard, 1975). The value of several of the larger estates was considerable.[8] By 1086, however, the region was suffering a period of economic depression, highlighted in Domesday Book by drastic falls in the values of manors.[9] The manor of Kilnsea, for example, was valued at £56 in the time of Edward the Confessor, but was worth only £6 in 1086. Although this depression was probably not as severe as in many other parts of Yorkshire, there being little evidence of large-scale population loss or land abandonment, it is likely that the new lord would have had a considerable interest in a rapid return to earlier levels of prosperity (Sawyer, 1969).[10]

Perhaps of even greater importance, however, was the tenurial complexity of the region in the eleventh century. The honour of Holderness had been created with little regard for pre-existing patterns, and as a result contained a complicated assortment of manors, berewicks and sokelands.[11] Not only would this make administration difficult, but also, under unified control, it was irrelevant. Could it be that the honorial lord took advantage of the depressed economic state of the region to reorganise and regroup existing holdings, perhaps creating larger, but fewer units (Wightman, 1975)? Field planning might then be seen as only one aspect of a major phase of reorganisation and adjustment affecting many facets of rural life.

While a strong case can thus be made for an origin of Holderness regular fields in the late eleventh or early twelfth centuries, it must be emphasised nevertheless that this argument rests entirely upon circumstantial evidence. Conditions favourable for planning may have existed at other times. Who, for instance, was the Viking nobleman or 'hold' from whom it is thought the area derives the first part of its name (Stenton, 1971, 509). Could a single lordship have been created following a late ninth-century Scandinavian conquest of the region, and if so, could field reorganisation have been instigated at this time? Recent contributions to the Scandinavian debate seem to favour the view that new, simpler systems of land division were introduced by the new settlers, resulting in an open-field system more regular and efficient than hitherto used (Finberg, 1972). So little is yet known of the details of these systems, however, that it would be dangerous to place too great a significance upon them at this stage. Furthermore, it must be questioned why planned layouts have not been found in areas such as East Anglia, where Scandinavian influence was even greater than in Yorkshire.

Another possibility might be that the form results from the initial settlement of the area. Some planned layouts in Europe undoubtedly were established at the time of inital colonisation and settlement, usually when that settlement process was conducted in an organised manner (Mayhew, 1973). In order for this to account for Holderness layouts, however, the area would have to have been settled at a time when open-field cultivation practices were already developed and in general use. Most scholars now believe that this could not have occurred before the eighth or ninth centuries at the earliest (Thirsk, 1964). Since both place-names and early tenurial structure seem to suggest that Holderness was well settled from early times, field planning must be considered to have been the result of reorganisation and not of initial colonisation and settlement (Jones, 1961a).

The Method of Planning

Evidence of Farm Holdings

If we accept that planned layouts are most likely to have been imposed upon an area which was already settled and under cultivation, it is important to consider how planning might have been undertaken. Were the new fields imposed with no account being taken of pre-existing arrangements, farm holdings, etc., or was some attempt made to equate these with the new divisions?

Research into similar field planning in continental Europe shows that considerable care was taken to relate old and new. For example, when the regular ordering of holdings, known as solskifte, was established in south and east Sweden in the early Middle Ages, it held the implicit understanding that each farmer's share in the new system should equal that in the old (Göransson, 1961). For simplicity and convenience the tax assessment of each settlement was used as a basis for land division, such that each share of that total tax was translated into the same percentage share of the total land area being divided. Measuring rods were used to achieve this, an agreed number of rods being equal to each unit of tax. There is even evidence that the total field area was determined with reference to the tax assessment, each unit of the assessment equating with a standard areal measure.

It is not easy to determine whether a similar practice was used in Holderness field planning. There are very definite indications, both in the post-medieval and medieval documentation, of a land division which resembled Scandinavian solskifte. In a number of townships there is evidence for the regular ordering of the farm holdings or ox-gangs through the fields, and in Preston this ordering can be reconstructed precisely, and even dated to about 1250 AD at the latest (Harvey, 1978). The strips comprising each oxgang had an identical relative location, and identical neighbouring strips, in all subdivisions of the two fields. Early medieval references to holdings in several townships 'lying towards the sun' or whose descriptions suggest standardisation of position in the fields seem to confirm the antiquity of holding regularity in Holderness.[12]

The main problem arises, however, when assessing whether the other, and more important aspect, proportional division, was also involved. The paucity of information regarding strip widths, together with the absence of maps showing open field structure, makes mensural analysis almost impossible. One potentially very significant point, however, is the use of the term 'bydale' in several Holderness townships. In Winestead and North Frodingham, for instance, specific areas of the open fields had the name 'Bydale'. In Preston, it had more general use. Each of the two arable fields was subdivided into seven bydales, and each of the oxgang holdings had a strip parcel in each bydale, that parcel always adjoining strips of the same neighbouring oxgangs.

The derivation of this name is of some interest, coming as it seems from OWScand., 'by', meaning 'a farmstead, a village' and ON. 'deille' meaning 'share, a portion of land'. It is tempting to conclude that the survival of this term provides evidence that a proportional sharing of land did occur in Holderness open fields. Perhaps the 'village share' or 'bydale' was a special part of the fields where individual shares were first measured out, and this then acted as a template for the division of the rest of the arable land.

Even if the fields were laid out in this way, however, there is still no evidence that the village tax assessment was used to achieve division, as was the case in Sweden. Where, for instance, a complete order of ox-gangs can be reconstructed, as in Preston, that order appears to have related to the actual number of holdings present in the township by 1250. It is difficult to see how this could have related to the tax assessment of the village recorded in Domesday Book. But we cannot be certain that the division of which we have evidence was necessarily the only division to have occurred. It is quite possible that several re-divisions of the fields took place, made necessary perhaps by population change or by a need to resolve complexities resulting from land exchanges. If, as has been suggested, field planning did originate before the mid-twelfth century, it is more than likely that the tax assessment would then have played a significant role, since at that time it was in regular use (Farrar, 1912).

There is therefore no direct means of assessing whether the internal division of the new fields was determined on a proportional basis. The problem can be examined indirectly, however. It has already been stated that in Scandinavian planning the area of the arable land was determined by the tax assessment, there being a clear proportional relationship between the size of a village's tax and the size of its fields. If it could be established that Holderness fields also varied in size with relation to their tax assessment, then we may be some way towards an understanding of how the fields were laid out.

Evidence of Township Areas

Data for such an analysis is readily available. Domesday Book records a tax assessment for almost all places in Holderness, and whilst this assessment almost certainly replaced an earlier one in the early eleventh century, it remained in regular use until the beginning of the thirteenth century (Sawyer, 1971, 172; Farrar, 1912). In common with other parts of Yorkshire, this assessment was given in terms of carucates and bovates. Information concerning field sizes is a little more difficult to obtain. The precise boundaries of the open fields are not always known, and even when they are, it cannot necessarily be assumed that they remained stable for long periods. Nor is it certain whether the assessment would have related only to arable land, or whether small areas of meadow might also have been included.[13] Because of these uncertainties it was decided to use the total township area as the basis for an initial analysis. This is obviously less accurate, but it can be justified not only because all townships can be examined, but also because in many cases the arable fields occupied all but a small part of this area. Major problems might only arise, therefore, in townships with extensive meadow lands.

Any possible relationship between area and tax can best be expressed statistically, using the correlation coefficient. When data for all townships was tested in this way, the correlation was +.39 (Figure 25).

Figure 25: A Graph to Show the Relationship between Township Area and Domesday Carucate Assessment

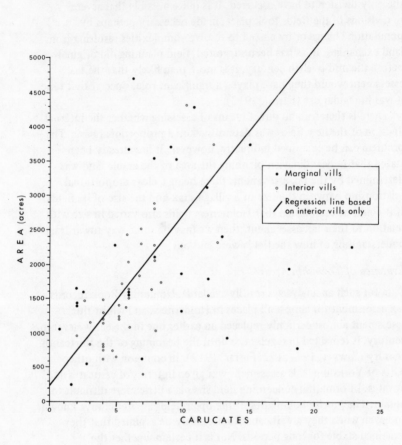

This proved to be significant at the 99 per cent level. Even so, there was a large number of townships where no relationship could be observed. Significantly, a majority of these fell into one of two groups. First, coastal townships, whose areas had been considerably reduced by marine erosion, and whose tax assessments were generally much larger than would be expected from their modern areas. Secondly, townships on the south and western margins of Holderness included extensive amounts of meadow land, and their assessments were usually far smaller than might be expected from their areas. When all these marginal townships were excluded from the analysis, the correlation coefficient of area and tax of the remaining townships was much higher − +.70. This suggests not only that a very strong statistical relationship exists between area and tax, but also that it is a relationship which is unlikely to have arisen through chance.

If field areas had indeed been delimited on a proportional basis, using the village tax assessments, it is to be expected that a standard areal unit would have been used to equate with each unit of tax. Since such a unit cannot be identified directly from township areas, the closest approximation of the relationship between the two sets of data must be obtained statistically, by calculation of the regression line for area against tax. Using data for non-marginal townships only, this suggests that one unit of tax, or each carucate of the assessment, was equivalent to an area of 240 acres, or a square 1078 yards × 1078 yards. A township assessed at two carucates might thus be expected to have a field area containing two carucate squares, or 480 acres, whilst one assessed at four carucates would contain four carucate squares or 960 acres.

The arrangement of these squares is also likely to have taken a regular form in a planned system, and even if internal divisions are not known, the field boundaries might be expected to show evidence of symmetry and regularity. An examination of the boundaries of non-marginal townships (which need not necessarily be field boundaries) shows that in some 39 per cent of cases, the distance between one or both pairs of boundaries closely fitted the width of the notional areas or multiples of that width (Figure 26). In a further 40 per cent of cases the fit was not so obvious, usually because the boundaries themselves were more irregular, but it was still possible to see how the carucate squares might have equated with the general shape of the townships. Even more significant was the fact that when the analysis was extended to marginal townships, many of those boundaries which were least affected by changes in land area also conformed to this pattern. In all, some 46 per cent of all townships had one or both pairs of boundaries

Figure 26: The Relationship between Township Areas and the Notional Carucate Areas

closely relating to the carucate squares, with a further 34 per cent showing a more general similarity. Only 20 per cent of townships appeared to demonstrate no relationship whatsoever.

It is, perhaps, too early fully to assess the significance of these results, especially since township areas only provide a very general guide to field sizes in some cases. Even so, the very fact that township boundaries were so often field boundaries in Holderness, and that the arable land occupied so large a proportion of township areas suggests that certain conclusions concerning the fields are justified, at least in very broad terms.

The strong correlation between tax assessment and area and the close fit between township boundaries and idealised carucate areas would certainly seem to provide supportive evidence for a hypothesis that Holderness planned fields were laid out on a proprotional basis, using the tax assessment as the means of determining the actual areas of the new fields. It would also follow that internal divisions of these fields, including the strips of individual farm holdings, are likely to have been measured out in a proportional way, again using the assessment. It must be emphasised, however, that no such causal linkage is in fact established by this investigation. Very similar results might arise from a situation in which the tax assessment was closely adjusted to an agrarian reality in which almost all potentially cultivable land had been utilised, and where township boundaries had already been delimited. In view of the supposed artificiality of the assessment itself this may be unlikely, but it is still worthy of future inquiry especially in the light of recent opinion concerning the relevance of the eleventh-century assessment in Yorkshire to agrarian conditions of that time (Round, 1895; Maxwell and Darby, 1962).

While, therefore, there may still be some debate concerning the interpretation of this relationship between tax and area, one fact does seem to have been established with some certainty. Regularity and planning are just as likely to have applied to township areas as to the internal layout of fields and the ordering of holdings. However, it is not possible at this stage to say whether all these features were established simultaneously, or whether in fact there was more than one phase of planning in the Holderness landscape. Perhaps this might be clarified by a brief examination of one other element in this landscape, the settlement pattern.

Evidence of Settlement Form

Holderness settlements are remarkable for their apparent lack of regularity (Figure 27). Analysis of late nineteenth-century Ordnance

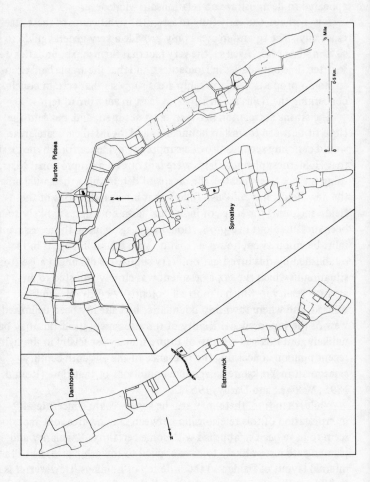

Figure 27: The Village Enclosures of Elstronwick, Danthorpe, Sproatley and Burton Pidsea

Survey maps shows that over two-thirds of villages were linear, and many were over one mile in length. Village enclosures were usually of irregular shape and size, and had no general alignment or frontage. It is also clear that some settlements were not simple in form, but included within their plans both linear and cluster elements. Thus it is possible to identify in the overall linear form of Sproatley three or even four small foci or clusters of enclosures. Such arrangements suggest that they are the product of a long period of evolution rather than of planning, and the different morphological elements may indicate different phases in this evolution.

These complex settlements show a marked resemblance to the 'polyfocal' forms identified by Taylor in Cambridgeshire and elsewhere — forms which he argues could have originated through the linkage of once separate nuclei (Taylor, 1973; 1977). These distinct village foci can sometimes be identified in the medieval period when they appear to have related to separate manorial holdings. In Holderness also, it is true that a majority of composite settlements were associated with more than one estate at the time of Domesday Book. Sproatley, for example, contained four different tenurial units. In contrast, places having a simple plan in the nineteenth century were almost invariably single holdings in 1086. It certainly looks as though a number of Holderness villages may well be secondary forms created through the linkage of once quite separate but neighbouring nuclei.

If small settlement nuclei were once more common in the region, then this could have important implications for the debate on field origins, for this reason. Both simple and polyfocal settlements were associated with the same planned field form. Does this mean that any separate nuclei which eventually became physically linked had, from an early period, shared in the cultivation of arable land common to them all, or has it other implications? For if even some of these nuclei had had their own arable land which they organised and cultivated independently of their neighbours, then a stage must be envisaged whereby this independence was broken down and separate arable areas became amalgamated (Finn, 1972). This might have occurred before the planned layouts were introduced, but it could not have taken place after that event.

In a similar way, the study of township areas also suggests that more than a nominal association existed between any independent settlements within them. For whereas each manorial holding was separately assessed, it was the total tax for all such holdings within any one township which formed the basis of the observed relationship between area and tax. In Sproatley, therefore, the township area related quite closely to the combined assessment of all four holdings, and not to any one manorial unit. Clearly, if at least some of these holdings had originally been separate both physically and economically, this could no longer have been the case by the time the field areas were delimited — assuming, that is, that the division was made on a proportional basis, using the tax assessment.

Although the study of both settlement form and early tenurial structure may provide supportive evidence of the role of reorganisation and amalgamation in the development of Holderness townships, it does not directly establish an association with field planning itself. Nevertheless, it would seem strange if amalgamation and planning were unconnected, since neither is liable to have arisen at a time of weak or fragmented lordship. Furthermore, if the role of the tax assessment has been appraised correctly, any planned reorganisation is more likely to have had a post-Conquest rather than pre-Conquest origin, for two reasons. First, the medieval assessment itself was probably only fixed in the early eleventh century (Sawyer, 1969), and secondly it is difficult to understand why each holding should have been separately assessed at this time unless they were still independent units. The decades following the Norman Conquest do indeed seem to have been crucial for the establishment of regularity and order in the Holderness landscape.

Conclusion

The conclusions reached in this paper are based primarily on surviving documentation and analysis using a geographical approach. Inevitably, since the origins of Holderness regular fields are undocumented, the investigation hinges upon speculation and extrapolation, and its results are still, therefore, a matter for debate. Any further study of the possible origins of Holderness, for example the dating of strips or of settlement amalgamation, would depend upon archaeological research. If nothing else, this study highlights the need for an interdisciplinary approach to landscape analysis.

Notes

1.   Humberside County Record Office. DDCC(2), G.2.
2.   A photograph of this map is in the possession of Captain Hildyard, Winestead Hall, Winestead, North Humberside.

3. The small area of short, east-west orientated strips in the east of South Field was also subdivided.

4. B M Cotton Otho, cviii, f79.

5. A number of documents describe details of Halsham field structure, in particular HCRO DDCC/App C (A Lay Cartulary), and HCRO DDCC/43/1-30 (a series of land grants).

6. Known documents of around this time all indicate the presence of longlands in the townships to which they relate.

7. Principal sources of evidence are several thirteenth- and fourteenth-century surveys of knights' fees (all printed in Surtees Society Publications LIX pt. 1, 1866), and Domesday Book.

8. For example, the manor of Burstwick, with its appendages, was valued at £56, that of Hornsea and Kilnsea also £56.

9. For instance the value of Hornsea fell from £56 TRE to £6 in 1086.

10. There is little evidence in Holderness of the wholesale destruction caused by William I's 'Harrying of the North' in the winter of 1069-70, but the region would inevitably have been adversely affected by the depression elsewhere. Scandinavian raids along the east coast were still taking place around this time, and they also may have had a detrimental effect on the area (see, for example P.H. Sawyer, 1969).

11. A sokeland was a large estate, consisting of a principal manor which had dependencies or sokes in several other places.

12. Göransson considers this phrase to be a clear indication of solskifte-like ordering.

13. Göransson indicates that the field areas delimited by the Scandinavian tax assessment included both arable and meadow land.

10 EARLY CUSTOMARY TENURES IN WALES AND OPEN-FIELD AGRICULTURE

Glanville R.J. Jones

In 1303, as a response to their petition to the Prince of Wales about the holding of escheated lands, the tenants of North Wales were assured that such lands should not be enclosed or otherwise separated than was customary in the time of the former tenants. These escheated lands were to be held 'openly and not otherwise (*aperte et non aliter*)' (Ellis, 1828, 213). Clearly therefore some at least of the agricultural lands of North Wales lay in open field. Yet, the north-western portion of North Wales had formed the nuclear area of the last independent Welsh kingdom of Gwynedd, which had been finally conquered by the English only in 1283. Thus, despite opinion received to the contrary, open-field agriculture was not confined to those parts of Wales which had first succumbed to Norman conquerors. On the contrary, throughout Welsh Wales (*Pura Wallia*) agriculture was practised on land subject to common rights and in large part lying physically open. This practice of open-field agriculture is clearly reflected in the main customary tenures of Wales as portrayed in medieval Welsh lawbooks and further illuminated in post-conquest extents or surveys undertaken in order to record rents and services attached to land. Nevertheless, these tenures also reveal that many features of territorial organisation were archaic, for the rural economy of medieval Wales, though one of great complexity, was underdeveloped by English standards. Consideration of the agrarian arrangements associated with these tenures should therefore help to illuminate early stages in the development of open-field agriculture in England.

Hereditary Land

The escheated lands which were the subject of the 1303 petition had formerly been subject to hereditary land tenure, the most widespread of all the customary tenures of Wales (Jones, 1972). The essential features of such hereditary land were that the rights to it passed to male descendants in equal shares, and that the rights of the 'owner' for the time being were limited to his lifetime so that he could not defeat the right of his own descendants to succeed. Continued occupation of land by the members of an agnatic lineage over a prescription period of four generations ensured the conversion of bare possession into legal proprietorship. By the thirteenth century a hereditary proprietor's share of his patrimony would typically have consisted of a personal holding of appropriated land and an undivided share of joint land (*cytir*). The appropiated land would have included a homestead (*tyddyn*), some long narrow parcels or strips of 'scattered land' (*gwasgardir*) lying intermixed in one or more open-field arable sharelands, and some plots of meadowland. The joint land would have comprised an expanse of wood, pasture and waste, with the latter often containing some bogland used as a turbary. To judge from the surveys of the late thirteenth and early fourteenth centuries each proprietor would have exercised within this joint land rights of common pasture and of temporary cultivation alike; these rights would have been estimated in terms of his acreage of appropriated land.

In accordance with the rules of partible succession (*cyfran*), the inheritance of a dead man was normally divided equally among his sons. By the thirteenth century, the appropriated land was usually divided physically; and, correspondingly, rights over the still undivided joint land were reduced in proportion. According to the Book of Iorwerth, the lawbook which relates particularly to Gwynedd, the intermixed parcels which made up the equal shares of appropriated land were to be chosen by the brothers in order of seniority (Wiliam, 1960). The youngest son was to inherit the father's homestead but the other homesteads were divided among the remaining brothers, again in order of seniority. In practice, however, these other brothers are likely to have remained in the homesteads they had each already made for themselves on the hereditary land in their father's lifetime; for a scion of free stock, on attaining fourteen years could settle, if he so wished, in his own homestead on part of the family land. The lawbooks which relate to other parts of Wales give slightly different rules for the inheritance of patrimonies. Nevertheless all versions have sufficient elements in common to show that hereditary-land tenure was widespread throughout Wales (Richards, 1954; Emanuel, 1967).

This division of the patrimony was to continue during the lives of the brothers. After they had died, however, their sons, being first cousins, could revise the sharing of the land if they so wished. Similarly, the second cousins had a right to a revision of the sharing. If only some cousins or second cousins so wished, they could compel their kinsmen to join in a re-sharing of their common great-grandfather's holding. But

there was no general obligation to re-open an arrangement made in a
previous generation for the sharing of land; and, in any event, re-sharing
was permitted only as far as the second cousins, otherwise described in
the Latin law-texts as the sons of cousins. Clearly therefore, the four-
generation group, extending vertically backwards as far as the great-
grandfather and horizontally outwards as far as the second cousin, was
important as marking the limits within which re-sharing was permitted.
Again, if a kinsman were to die without leaving a descendant then his
brother, or cousin, or second cousin, could claim his land, if a common
ancestor had held it. Moreover the hereditary land did not finally
escheat to the king, or in the words of the lawbooks become
extinguished, if a third cousin were alive to claim it.

In Wales it is probable that on the land recognised in the lawbooks as
hereditary there were four-generation inheritance groups like those of
early Ireland. In Ireland before the seventh century rights over
hereditary land were shared equally among the adult male members of
the four-generation group and re-allocation occurred with the passage
of each generation (Charles-Edwards, 1972). That the same had once
been true of Wales is suggested by the rule in Welsh law that the re-
sharing of hereditary land did not extend beyond second cousins.
Indeed one south-Welsh lawbook when dealing with the sharing of
patrimony refers to the land of the four-generation group as the joint
land (*tir cyd*) on which 'beyond second cousins no one is to preserve
the share (*ran*) of another'. The use, in this context, of the word 'share'
supports the suggestion, advanced by Charles-Edwards, that the four-
generation group of Wales, like that of Ireland, was responsible at an
earlier date for providing each of its adult members with a standard
holding (Wade-Evans, 1909; Charles-Edwards, 1971). In Wales this
standard holding was at first probably similar to the unit named in a
number of South-Welsh lawbooks as a shareland (*rhandir*) of 312 'acres'.
Within such a shareland 12 'acres' were said to be for building and 300
'acres' were for arable, pasture and fuel wood. These 'acres' varied in size
from one locality to another but they were all of very small dimensions
so that the 300 'acres' of the shareland would have ranged in statute
measure from as little as 31.73 acres to 67.50 acres. With the growth of
population the standard holding of some lineages would have been
reduced in size. In other cases however it was the lineage itself which
would have been reduced in size: hence the situation described in the
lawbooks in which there might be on the joint land only one inheritor
who was therefore to have the whole of the land.

After about 1100 AD sharing by the four-generation agnatic lineage came to be displaced by equal division among brothers as the normal inheritance custom. As a result, the rule that a kindred must split up with each new generation also appears to have disappeared. Consequently, the agnatic lineage expanded from one generation to the next and became a large kindred known as a *gwely*.

The term *gwely* had hitherto been used of a defined group of relatives within a wider circle of kindred responsible for payments of compensation in lieu of blood-feud. The same word *gwely* which also means 'bed' (*lectus*) came to be used of the landholding of the kindred in the post-conquest surveys of Wales. Thus, for example, at Little Dinorben in Denbighshire one *lectus* held by freemen and called Gwely Gruffydd ap Trahaearn was said in 1334 to contain half the total area of the vill, namely some 103 statute acres (Vinogradoff and Morgan, 1914). Again, near Bangor in 1306 a group of eleven tenants held freely in one *lectus* eleven messuages and twenty bovates of land (Ellis, 1828, 95); and in the Anglesey vill of Carwed in 1352 one free *gwely* was said to contain one carucate of land (Ellis, 1828, 73). Clearly therefore, as is implied by such expressions as 'the Gwely of Gwely Hebogyddion' used of a *gwely* in the Caernarvonshire vill of Dinlle in 1352, *gwely* had come to mean the stake in the soil or literally the territorial 'resting place' for the expanded lineage, in this instance that of the Hebogyddion (Falconers) (Ellis, 1828, 23; Jenkins, 1967).

Originally the cropland of a four-generation lineage would have included some permanent sharelands which, in whole or in part, were cultivated every year, and other temporary sharelands which were only taken from the common pastures and cultivated when needed. The permanent sharelands which adjoined the old settlement (*hendref*) of the lineage formed what may be regarded as infields; the temporary sharelands, on the other hand, were more distant from this nucleus of settlement and, alike in the lowlands and the uplands, formed what may be regarded as outfields in the midst of the common pastures. Particularly when located in the uplands the temporary arable was known as 'mountain land (*terra montana*)' (Jones, 1964). Such land was usually cultivated when the surrounding common pastures were grazed in the summer months, for near at hand would be sited the temporary summer dwellings of the members of the lineage.

The temporary sharelands of any expanding lineage would soon have become permanent if physical circumstances were suitable, and would thus have attracted permanent settlement. In the interests of economising on scarce resources of arable land any permanent dwellings established away from the old settlement (*hendref*) were usually sited on the

206   *Early Customary Tenures in Wales and Open-field Agriculture*

peripheries of the arable sharelands and thus formed girdle patterns of homesteads with each girdle flexibly adapted to the contour and thus to the lie of the land.

There were obvious physical as well as juridical limits to the territorial expansion of any lineage. Once this stage was attained any continued increase in the number of heirs would have brought about a progressive fragmentation of arable holdings and thus of proportional rights of pasture. In due course however some individuals, particularly those endowed with extraneous resources, were able to acquire the more fragmented hereditary holdings of less fortunate heirs and, by adding strip to strip, were gradually able to establish consolidated blocks of arable land to which were added adjoining portions of meadow and parts of the common pastures. Thus it was that enclosed fields came to replace the formerly open sharelands, as also the meadows and the unenclosed common pastures originally associated with hereditary-land tenure.

In medieval Wales hereditary land was not confined to freemen but could also be held by bondmen. How this could have arisen is indicated in those sections of the lawbooks which deal with aliens (Wiliam, 1960, 58-9). Thus aliens were said to become proprietors in the fourth generation after they had been settled on the lord's waste. Similarly the aliens of notables became proprietors in the fourth generation if they had occupied land under these notables for so long a time. Thenceforward such aliens were bound to the soil in bondage. After an alien had become a proprietor, he was to inherit a house with adjoining land, literally a homestead (*tyddyn*), and also arable land lying within sharelands. In other words, although now bond, he held land like the freeman by hereditary-land tenure. Since the alien appears to have acquired his own house and house croft only in the fourth generation he must have lived on another kind of land before attaining the status of proprietor. The nature of this land can be discerned from the rules in the lawbooks about a second customary tenure known as nucleal-land tenure.

Nucleal Land

Whereas on hereditary land each brother acquired his own house and house croft, nucleal land was to be shared as gardens, and not as homesteads. Moreover, if there were buildings thereon the youngest son was no more entitled to them than the eldest, but they were to be shared as cells or rooms (*ystefyll*) (Wiliam, 1960, 58). Nucleal land therefore appears to have been a variant of hereditary land. Gardens (*gerddi*) of nucleal land appear to have been arranged in a radial fashion around a nucleus of some kind, such as that containing a church. Thus the gardens which appear to have radiated from the enclosure containing the church were said to be a legal acre in length with their end to the churchyard (Wiliam, 1960, 44). These gardens were probably encompassed by a strong fence so that they could not be entered by beasts. Since these gardens were regularly manured they could be cultivated year in and year out like an infield but, in addition, the tenants would also be permitted to cultivate outfields.

In the name *corddlan* the first element *cordd* could refer to a group of some kind, probably a kindred or even a herd of cattle. The second element, *llan*, possibly before it came to be applied to a church in its churchyard, referred to an enclosure. *Corddlan* therefore could have meant the enclosure of a kindred. In this connection it is significant that according to the lawbooks, an orchard was one of the ornaments of a kindred (*cenedl*), the others being a mill and a weir. Such ornaments of a kindred were 'not to be divided nor removed but their produce shared between those who may have a right to it' (Wiliam, 1960, 58).

The occupants of nucleal land appear to have been under-tenants. They sometimes held land under notable proprietors belonging to a kindred, but more frequently they were under-tenants to a *clas* or church community. The title of the chief officer of the *clas*, the abbot, and the way in which the community shared the revenues of the church, indicate that the *clas* was at first a monastery. Later the *clas* became a community of canons with each canon having a share of the lands. Certainly by the time the oldest extant texts of Welsh law were written in the early thirteenth century the men of the community no longer took monastic vows such as those of celibacy. Indeed the normal *clas* had become a hereditary ecclesiastical corporation normally consisting of one or more kindreds in possession of the church and its lands, save that one portion of land was usually reserved for the maintenance of a priest. Nevertheless the privileges of such churches were jealously preserved so that they could offer sanctuary. Thus 'whoever shall take protection, his cattle are to be with the cattle of the *clas* and the abbots to the furthest limits to which they go, and afterwards are to be returned to their cattle pen' (Wiliam, 1960, 44). The under-tenants of such a *clas*, subject as they were to vows of chastity, would not originally have had heirs, so that their occupancy of nucleal land could not mature into proprietorship. When, however, church land came to be held by under-tenants subject to less strict rules, successive occupation over four generations would have enabled the occupant in the fourth generation to become a proprietor, and to acquire his own house.

Reckoned Land

A third tenure recorded in the lawbooks was that known as reckoned land (*tir cyfrif*) or geldable land (*tir cyllidus*). This was held by villeins, men without pedigree, whose shares of land were determined by the king's officers. According to the lawbooks these officers, the great reeve and the royal bailiff, allocated 'to every one in the vill (*tref*) as good as to each other' (Wiliam, 1960, 54), that is equal shares of land. Occasionally in the lawbooks, and more frequently in the surveys, the expression used for this tenure is *tref gyfrif* (reckoned township or vill), thus confirming that the land of the whole vill was shared equally among the bondmen. In return, these bondmen owed equal renders, although, as later case law reveals, the shares of land were equal in area rather than in value. When a bondman died, the great reeve and the royal bailiff were to correct the sharing if this were necessary. A revision of the sharing also took place when any, save the youngest, of a bondman's sons came of age at fourteen years. In other words, with any changes in the numbers of those having a right to land in the reckoned vill, the shares were reckoned anew, and the sons of a bondman holding reckoned land shared with his co-occupiers irrespective of relationship. Since reckoned land was not shared among brothers, there could be no extinguished acre in it to revert to the lord as escheat; if however an extinguished acre should lie within reckoned land, then the great reeve and royal bailiff should share it equally 'in common (*yn gyffredin*)' (William, 1960, 54). Despite the re-sharing of reckoned land which took place when necessary, a limited right to a particular house plot appears to have been recognised. Nevertheless when necessary a shifting of homesteads could take place in order to accommodate additional bondmen.

There was one reckoned vill in every neighbourhood (*cwmwd*) which was deemed to be of particular importance. This was the reeve's vill (*maerdref*) which adjoined the king's court. Just as the great reeve and the royal bailiff allocated shares to the men of the ordinary reckoned vill, so too the lesser reeve shared the land of the reeve's vill among the occupants; and, as far as possible, he too was to allow every bond tenant to retain his existing homestead.

The characteristic settlement of bond tenants on reckoned land was a hamlet. The complement of a legal hamlet was given as 'nine buildings and one plough and one kiln and one churn, and one cock and one bull and one neat-herd' (Owen, 1841, 692-3). It is presumably a communal organisation which is implied by this complement, hence the one plough, and the one churn, envisaged as serving the needs of nine homesteads. The attribution of one bull and one neat-herd to the hamlet likewise implies a communal organisation for the grazing of a common pasture. Hence the rule that 'if a bondman keeps a bull and it fails when required by a neighbour, the owner is subject to a fine' (Owen, 1841, 605-6); apparently the right of an individual to keep a bull was matched by an obligation to serve the cows of his neighbours. Such bond communities were moreover responsible for joint obligations to the king. Thus the cheese of their summer food gift to the king was made from milk produced at one milking of the cows possessed by all in the vill. Again each bond vill rendered for the king's hawks one ewe that had recently given birth. They also jointly supported groups of retainers on circuit (*cylch*) and provided the huntsmen, the hawksmen and the grooms with quarters (*dofreth*).

The Prevalence of Open Fields

The accounts of these three customary tenures as presented by medieval lawyers suggest that a form of open-field agriculture was prevalent in *Pura Wallia* in the Middle Ages. The numerous provisions of Welsh land-law about compensation for crop damage caused by grazing animals are barely comprehensible save in terms of open-field husbandry. No compensation however was to be paid for corn left standing later than the first of November, when the flocks and herds should have returned from their summer and autumn pastures. Thereafter, until 1 May when the animals departed for the summer pastures, most arable sharelands were grazed in common. Given this sequence the main crops cultivated were the spring cereals, oats and barley. Nevertheless with the aid of temporary fencing some areas in favoured localities were used for the winter cereals, wheat and rye. In addition the grass crop on meadowland was enclosed within fences from 17 March until 1 November. Confirmation that arable land had traditionally lain in open field is provided by statements about the rights of the beadle whose duties included that of acquiring for the king the moveable goods of any man who died intestate or without heirs; for from the 'house of death' the beadle himself was to have various things including 'the green flax, the lowest layer of corn unreaped in the earth and, if there be no headland, the skirts' (Wiliam, 1960, 19). Evidently the arable land normally lay unenclosed, bounded only by turf balks, but interrupted here and there by the occasional headland on which the plough was turned.

Open Fields as Recorded in Practice

The general picture of the agrarian economy of *Pura Wallia* presented by medieval lawyers was no doubt over-schematised. Nevertheless, that this picture was not too divorced from reality is amply demonstrated by the agrarian dispositions in the vicinity of a substantial number of well-documented medieval settlements.

Llanynys

At Llanynys, in the Vale of Clwyd, vestiges of both hereditary land and of nucleal land survived until recently and here, therefore, it is possible to gain a clear impression of their layout (Figure 28).

To this day one garden of former nucleal land flanks the road from Llanynys to Trefechan. It thus radiates north-westwards from the graveyard, which itself formed part of a larger and roughly circular enclosure. This was probably the original *llan*, for the field which adjoins the garden on the east was known in 1841 as Cae'r Llan (Field of the Church Enclosure) (Jones, 1977; Eidt, Singh and Singh, 1977).[1] The garden itself, though known as a croft in 1841, was earlier called Clwtt yn y Gerddi Duon (Piece in the Black Gardens), an appropriate name for constant manuring has long since darkened its naturally reddish-brown sandy loam soil. The use of the name Croft for four other fields immediately adjoining the church enclosure suggests that here too there had once been radial gardens. In addition, towards the north-west there had earlier been an outer group of radial gardens, and as late as 1808 one of these 'extended 119 yards to a stone in the ground' within the field called Bryn Castell (Castle Hill).[2] This outer garden, a headland at the southern edge of Bryn Castell, together with the Piece in the Black Gardens, and two of the crofts radiating eastwards from the church enclosure, had all retained their ancient dispositions probably because they were glebe lands for the support of the parish incumbent. Their status as glebe lands in itself testifies to their antiquity.

Beyond Bryn Castell on either side of the road from Llanynys to Trefechan the two fields called Maes Ucha (Upper Open Field) and

Figure 28: The Sites Occupied by Nucleal Lands and Hereditary Lands at Llanynys in the Vale of Clwyd

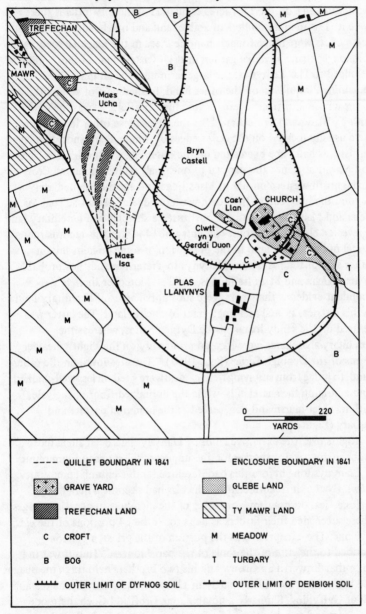

- - - - QUILLET BOUNDARY IN 1841	——— ENCLOSURE BOUNDARY IN 1841
GRAVE YARD	GLEBE LAND
TREFECHAN LAND	TY MAWR LAND
C CROFT	M MEADOW
B BOG	T TURBARY
╈╈╈ OUTER LIMIT OF DYFNOG SOIL	╈╈ OUTER LIMIT OF DENBIGH SOIL

Maes Isa (Lower Open Field) were until very recently subdivided into a
number of quillets or strips of arable bounded only by turf balks.
Earlier in 1841 on the 15¼ acres contained in these two fields there
were no less than 21 quillets of arable land and one headland. Most of
the quillets were much longer than the Piece in the Black Gardens, and
also longer than the radial garden in Bryn Castell. One of these quillets
belonged to Plas Llanynys, the large farm near the church. The
remaining 20 quillets, on the other hand, formed part of seven other
farms whose farmsteads were located further afield, as was the case
with Ty Mawr and Trefechan. Like the latter these other farmsteads
were usually located on the outer edges of former sharelands. Similar
farmsteads had once been sited on the outer edges of both Maes Ucha
and Maes Isa. Hence the survival of one small enclosure named Croft
at the north-western corner of Maes Ucha and another immediately
beyond the bye-road at the north-western edge of Maes Isa. That Maes
Ucha and Maes Isa had earlier constituted a shareland of hereditary land
is revealed clearly by a court-roll entry of 1444; for this records that a
parcel of hereditary land in the possession of some co-heirs lay on
either side of the road from Llanynys to Trefechan and thus probably
In Maes Ucha and Maes Isa (Jones, 1972).[3] Moreover there is
abundant evidence that there were substantial areas of hereditary land
in this district, as well as a small tract of nucleal land. The latter was
centred on the sandy loams of the Dyfnog Series, whereas the
hereditary sharelands were located principally on the slightly heavier
loams of the Denbigh Series (Ball, 1960).[4] The meadows, on the other
hand, to judge from surviving field names, were sited on less well-drained
land adjoining the sharelands, while the ill-drained floor of the vale,
with its alder marsh and bog, served as the common pasture and
turbary (Figure 28).

Rights over the extensive lands of Llanynys had been alienated
probably by royal authority for the support of the monastic community
of Llanynys long before this had developed into a church community
(*clas*). By the thirteenth century the *clas* had become a hereditary
ecclesiastical corporation consisting of the members of agnatic lineages.
These succeeded their fathers as heirs to all the 24 portions of the *clas*
save one. The exception was the portion of the priest, a portion
ascribed to the cure of the souls of the parishioners.[5] Thus it was in
1324 that Iorwerth ap Gronw and his two brothers held three messuages
and half a carucate of land partly in Llanynys and partly in the distant
vill of Gyffylliog. Similarly a notable, one Gruffydd Goch, held one
messuage and two bovates of land in Llanynys and Gyffylliog; but in

addition he also possessed in Llanynys some smallholdings which were
specified as '11 acres 3 curtilages' (Jack, 1968).[6] These smallholdings
were probably made up of ancient nucleal land which formerly had
been worked by the under-tenants of the monastery, but were now held
by the under-tenants of Gruffydd Goch. Such former nucleal land, thus
transformed into hereditary land, was subsequently to undergo the
same process of change as that which was to transform most of the
unenclosed quillets in the hereditary sharelands into enclosed fields.
Only the quillets of glebe in the portion of the priest escaped this
process and thus survived long enough to reveal the characteristic
disposition of ancient nucleal lands.

Eglwys Ail

Around the church of Llangadwaladr in the Anglesey township of
Eglwys Ail the ancient dispositions of nucleal land remained even more
evident until the eighteenth century. Thus here in 1747 there were a
number of small quillets which, in the words of a contemporary record
were 'bounding on the church-yard'.[7] Some of these were quillets of
glebe land but the remainder belonged to a number of prominent
landowners (Figure 29). These quillets radiated outwards from three
sides of the rectangular church-yard, but it is likely that they had once
radiated from all sides of the church-yard as is indicated by the layout
of the enclosures on its western side. The siting of a dwelling at the
outer limit of one quillet on the eastern side and of another dwelling
at the outer limit of an enclosure on the western side suggests how the
dwellings on nucleal land at Llangadwaladr had earlier been arranged in
relation to the radial quillets or gardens. Again, the latter were much
smaller than other outlying quillets in the area of the township, like
that, for example, which cut across Cae'r Stent (Extent Field). This
contrast may in itself be significant for the name Stent (Extent)
indicates that this was escheat land which, because it had escheated to
the Crown, was subject to an increased or extended rent. As such, this
particular outlying quillet had undoubtedly formed part of that
portion of the hereditary lands of Gwely Gwas Sanffraid ap Tanharn
which had become escheat by 1352. This *gwely* was one of the two
gwelyau in the township of Eglwys Ail, a township which itself was
named after a wattled church. The township was described in a survey
of 1352 as being 'free and held of St. Cadwaladr the king' (Ellis, 1828,
46-7). By this tenure the 'heirs' of the two *gwelyau* rendered nothing
for their lands save that certain rights were reserved to the Crown. As
the name of this tenure implies, and early fourteenth century accounts

Figure 29: The Dispositions of Former Nucleal Lands and Hereditary Lands at Llangadwaladr, Anglesey

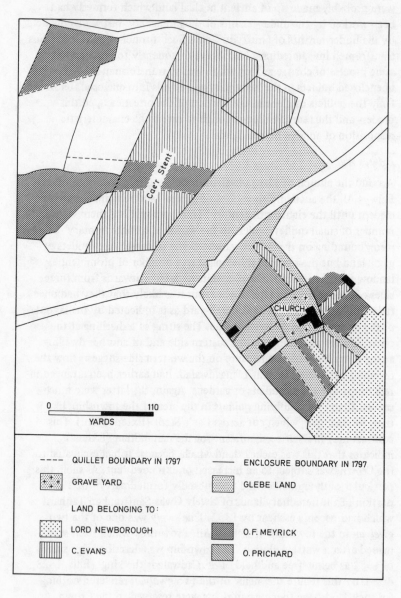

confirm, these were lands held *de sanctuario*, that is as sanctuary lands which were exempted from the payment of ordinary rent.[8] The Cadwaladr whose name is commemorated in that of the tenure, as a king and a saint alike, was a seventh-century ruler of Gwynedd who, according to tradition, became a monk before the end of his days. The monastery to which he retired was undoubtedly that on the site of the church of Eglwys Ail which was re-named Llangadwaladr after him; for at this church is an early Christian inscribed stone commemorating the grandfather of Cadwaladr, the renowed king Catamanus (Cadfan) who had died by the late 620s (Nash-Williams, 1950). Certainly the name given to the tenure of the lands of Eglwys Ail suggests that Cadwaladr either endowed the community at Eglwys Ail with these lands or permitted these lands which were already in the possession of the community to be held on more privileged terms. It is likely therefore that at Llangadwaladr from an early date there were nucleal lands held by under-tenants of the monastic community. By the thirteenth century the lands of this community had come into the possession of the heirs of the two *gwelyau* in the township of Eglwys Ail. Nevertheless it is likely that there were still some nucleal lands held by the under-tenants of these heirs, even in the fourteenth century, for in 1352 the heirs could claim that they were 'free to mill in their own houses' (Ellis, 1828, 46-7; Emanuel, 1967).[9] Presumably therefore they had under-tenants numerous enough to deal by means of hand querns with all the milling of their grain. In other words, the evidence available for the medieval township of Eglwys Ail suggests that here in the fourteenth century, besides the lands held by hereditary-land tenure, there were also lands held by nucleal-land tenure.

St Asaph

At Llanelwy, or St Asaph as it came to be known, the medieval tenurial dispositions were even more complex. Here in 1143 a bishop's seat had been established at the site of an ancient *clas*. Near the cathedral church of Llanelwy, as it was described in 1188, stood the medieval palace or court which housed not only the Bishop of St Asaph but also his substantial household. For the sustenance of this court there was at Llanelwy a substantial area of arable demesne land which had originally been cultivated by the labour of the Bishop's tenants. The Bishop's estate of Llanelwy appears to have extended into the four subsidiary townships of Talar, Gwerneigron, Gwernglefryd and Brynpolyn into which Llanelwy was divided (Figure 30). No less than 90 acres of this arable demesne were deemed to be of the best quality and presumably

Figure 30: The Dispositions of Demesne Lands and Hereditary Lands in the Estate of the Bishop of St Asaph at Llanelwy in the Vale of Clwyd

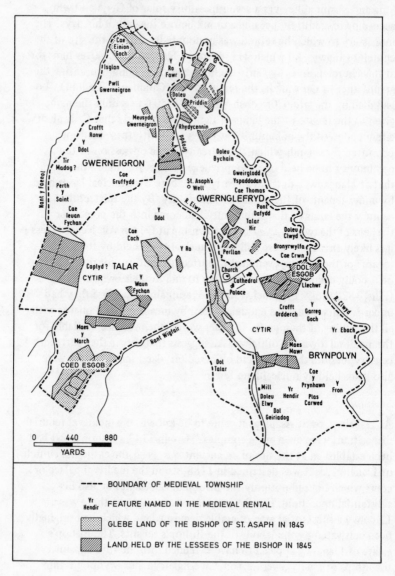

----	BOUNDARY OF MEDIEVAL TOWNSHIP
Yr Hendir	FEATURE NAMED IN THE MEDIEVAL RENTAL
	GLEBE LAND OF THE BISHOP OF ST. ASAPH IN 1845
	LAND HELD BY THE LESSEES OF THE BISHOP IN 1845

therefore cultivated regularly.[10] There were, in addition, 120 acres of poorer land which were probably cultivated only intermittently. By the fourteenth century these arable demesne lands had been farmed out, but nevertheless they still lay intermingled with the lands of the Bishop's tenants. Among these, according to a detailed rental of the late fourteenth century,[11] there were eight bond tenants who performed works at the church of Llanelwy. Far more numerous however were the occupants of the seven *lecta* (*gwelyau*) in the four subsidiary townships of Llanelwy. Each *lectus* was occupied by the descendants of the *gwely* eponym, but in most of the *gwelyau* these descendants were subdivided into a number of progeny. Thus there were ten progenies in Lectus Aldryd and three progenies in Lectus Kedmor. To add to the complexity of these tenurial arrangements some of these progeny in turn could have under-tenants. Such, for example, was the case with two of the five progeny in Lectus Uliar.

The rental for this estate is unfortunately incomplete but, in those sections which survive, five *gwelyau* were surveyed in full. These five *lecta* were occupied by at least 82 proprietors of hereditary land. Of these 82 proprietors no less than 34 held, by Welsh vifgage, additional parcels of hereditary land which had formerly been in the possession of other proprietors. In addition, there were at least 30 other occupants who held hereditary land by Welsh vifgage but who held no other hereditary land by hereditary right. When account is taken of the two *gwelyau* not surveyed in the rental there must have been at least 160 family households in Llanelwy during the late fourteenth century, apart from the occupants of the Bishop's court. From an area of 2,043 acres contained within the four subsidiary townships, these family households eked out a living from the lands not included within the Bishop's demesne. The holdings of proprietors and other occupants of the five *gwelyau* varied very considerably in size. The typical holding of such hereditary land was however not only small but also fragmented. The dispositions of most of the parcels of hereditary land recorded in the rental can be ascertained from their description in relation to features whose names have survived (Figure 30).

By the late fourteenth century the arable lands of some of the *gwelyau* extended almost to the very limits of Llanelwy as was the case, for example, with a holding in Lectus Uliar described as lying near the junction of the rivers Clwyd and Elwy (Figure 30). All five *lecta* embraced lands in each of the four subsidiary townships of Llanelwy but the lands in most localities formed part of only one *gwely* or, at best, two. Thus the lands in Croft Ronw in the township of Gwerneigron

formed part only of Lectus Aldrid, whereas the lands of Y Ddol, said
to be near Croft Ronw, formed part only of Lectus Uliar. On the other
hand, Bronyrwylfa in Gwernglefryd formed part of both Lectus Aldrid
and of Lectus Kedmor. By contrast, there were parcels of land belonging
to all five *lecta* at Llanelwy proper, the settlement adjoining the
cathedral in the north-west corner of Brynpolyn township. Given the
representation in a small area here of all five *lecta* these parcels at
Llanelwy proper were bound to be small. Such was the case with a
parcel in Lectus Segynnabiaid, said to be within the vill of Llanelwy
near the cemetery of St Asaph, in other words near the parish church of
Llanelwy. This parcel contained only 1½ rods and although the acre
used in the rental was a customary unit containing 10,240 sq. yds, this
parcel would have contained in statute measure only 96 sq. yds. Yet
this tiny parcel was held jointly by the Dean of St Asaph and
Gwenllian, the daughter of Einion Bleddyn. By contrast the
consolidated parcel of land, meadow and wood, which the Dean and
Gwenllian likewise held jointly in Bronyrwylfa near the River Clwyd
contained no less than 16 customary acres (33.85 statute acres), and
the lands they held at Yr Ebach contained a further 4 customary acres
(8.46 statute acres). Again a parcel of hereditary land in Lectus Aldrid,
said to be in Llanelwy between the manor (i.e. the palace of the Bishop)
and the house of Cynan, contained only 6 rods (384 statute sq. yds).

Many of the small parcels in Llanelwy proper and notably those
called *places* were recorded in the rental as the sites of houses. Some of
the small parcels were called gardens or orchards; and seemingly just as
the Bishop had a garden and an orchard in Llanelwy proper so too did
some, at least, of his tenants. Among these gardens was Gardd y Perllan
(Garden of the Orchard) in Lectus Uliar where Gwenllian and
Tangwystyl, the 'daughters of the Orchard *(filie y Perllan)*' held a parcel
of 24 rods (1,536 statute sq. yds). In addition these two sisters held, by
Welsh vifgage, a parcel of 54 rods (3,456 statute sq. yds) of the land of
Michael y Gof (the Smith), said to be 'within the Garden of the
Orchard *(infra Gardd y Perllan)*'. Clearly therefore gardens could
contain small parcels of land which had formerly belonged to different
proprietors, a situation reminiscent of the legal provisions for nucleal
land. This parcel in the Garden of the Orchard was not the only
hereditary land that Gwellian and Tangwystyl held within Llanelwy
proper; there too they held an even smaller parcel of hereditary land
containing only 10 rods (640 statute sq. yds). Although the total area
of hereditary land held by the two sisters was only 3 acres 57 rods (7.10
statute acres), their hereditary lands lay in at least four separate places.

Thus they held, by hereditary tenure, 1 acre 24 rods (2.43 statute acres) near Perth y Saint over a mile to the west of Llanelwy proper and 'in the same open field (*in eodem campo*)' they held another parcel containing 42 rods (2,688 statute sq. yds). Again they held, by hereditary tenure, ¾ acre (1.59 statute acres) on the border of Gwernglefryd near the river Elwy (Figure 30). As would be expected, these outlying parcels held by the sisters were larger than those of their parcels which lay in Llanelwy proper.

Similarly Meredydd ap Madog ap Cynwrig held hereditarily in Lectus Segynnabiaid 'within the vill of Llanelwy (*infra villam de Llanelwey*),' that is in Llanelwy proper, an area of 18 rods (1,152 statute sq. yds) with one house. There, in addition, between the manor of the Bishop and the house of Cynan, he held a parcel of 32 rods (2,148 statute sq. yds). Further afield he held much larger parcels. Thus in one *place* with one house and in two parcels annexed to his house in Brynpolyn he held hereditarily $8\frac{7}{8}$ acres of land (18.79 statute acres). Near this house, too, at Doleu Elwy he held 2 acres 72 rods (5.20 statute acres); and on the river bank near this house a further 4¼ acres (8.99 statute acres). Close at hand near the mill were diverse parcels containing 4 acres 8 rods (8.56 statute acres) and near Dol Geiriadog diverse parcels containing $1\frac{1}{8}$ acres (2.38 statute acres). Clearly therefore the more substantial holdings of hereditary land on the Llanelwy estate were made up of tiny parcels at the core of Llanelwy as well as larger outlying parcels. According to the rental, a proprietor's homestead was normally located in the outlying locality but in addition there were houses or buildings on the smaller parcels at the core of Llanelwy. Thus, for example, the messuage of Gruffydd ap Robyn, one of the largest proprietors in Lectus Aldrid, was in Talar near Cae Coch (Red Field), but in addition he had in Llanelwy proper in his *place* 48 rods (3,072 statute sq. yds) 'with his houses and buildings (*cum domibus et edificiis suis*)'. Since there were numbers of houses and buildings on some core parcels it is likely that these were nucleal lands occupied by under-tenants.

The presence at the old church settlement of Llanelwy of nucleal lands, some of which were associated with orchards or gardens, recalls the statement in the Book of Iorwerth that the three ornaments of a kindred were an orchard, a mill and a weir. Later case law, however, records that an orchard, a mill and a weir were to be 'three erections to be common among brothers' (Owen, 1841, 687-8). This suggests that originally the ornaments of a kindred should be ascribed to the period before sharing by four-generation inheritance groups came to be displaced by equal division among brothers. Certainly at Llanelwy the former presence of four-generation inheritance groups is suggested by the naming of one parcel in Lectus Aldrid as Tir Mab Y Cefnder (Land of the Son of the Cousin).

Such older tenurial arrangements appear to have persisted in actual practice in one locality. This was Y Groydd, recorded in the rental as containing 3 acres 50 rods (7 statute acres), lying on either side of the River Elwy in Gwernglefryd near Rhydycennin (Figure 30). The name Groydd — a plural form of Ro — clearly refers to the pebbly strands or beds of the River Elwy. Y Groydd can therefore be identified positively as the area south of Rhydycennin near the western limit of Gwernglefryd where the Elwy bifurcated into a number of channels and thus diverged in part from the township boundary. Accordingly Y Groydd could be said to lie literally on either side of the Elwy. A thirteenth-century lawbook records that whoever possessed land upon the margin of the shore possessed 'as much of the beach as the breadth of his land' and could make 'a weir or other things thereon' (Wiliam, 1960, 58). The pebbly strands of Y Groydd and the adjoining lands were however in the hereditary tenure of a group of tenants in Lectus Uliar including two cousins, an uncle and three others who, at best, were only second cousins if not yet more distantly related. It would appear therefore that Y Groydd provided the site of a weir, presumably of the kind said in the lawbooks to be lawful for catching only fish (Wiliam, 1960, 101-2). This weir was probably still the ornament of the kindred made up of the occupants of Lectus Uliar. If so, its late survival would lend support to an interpretation of the orchards at Llanelwy proper as having earlier been the ornaments of a number of kindreds. Unlike the weir of Y Groydd, whose lands were not divided, these orchard ornaments had been subdivided into tiny parcels of nucleal land. These, to judge from the later dispositions of parcels ascribed to the Dean and two of the prebendaries of St Asaph, were disposed radially in relation to the old parish church at Llanelwy,[12] a structure which must have pre-dated the first cathedral of St Asaph. From the same fourteenth-century rental, hints too can be gleaned of the presence of similar small parcels of nucleal land in the vicinity of the well of St Asaph some half a mile to the north of the parish church (Figure 30).

Analysis of the Llanelwy rental suggests therefore that even in the fourteenth century kindreds could possess as ornaments a number of spatially separate nuclei. Hence perhaps the statement in one lawbook that every habitation — presumably every outlying homestead on hereditary land — ought to have two footpaths, one to its church and one to its watering place (Owen, 1841, 269-70). On the Llanelwy estate

in the fourteenth century such footpaths would have served to
articulate the tenurial system by providing the necessary links between
outlying homesteads on hereditary land and any ornaments of a kindred
which had survived, whether in whole like Y Groydd, or only in part
like the orchards near the parish church.

Parcels of nucleal land notably those near the parish church of St
Asaph, and probably also those near the well of St Asaph, though
fragmented, would have been used as the sites of buildings from an early
date and, as a result, enclosed. On the other hand, even the parcels of
arable and meadow near outlying homesteads on hereditary land were
frequently unenclosed. Cae Coch (Red Field), part of Lectus Aldrid in
Talar, was near the messuage of Gruffydd ap Robyn. It contained $9\frac{1}{8}$
acres (19.31 statute acres), nearly one quarter of Gruffydd's holding,
which was itself in the process of being consolidated. Nevertheless,
Gruffydd's land in Cae Coch was still described in the rental as being 'in
the same field within boundary stones and fences near his barn (*in eodem
campo infra metas et sepes iuxta horreum suum*)', so that part of it must
still have lain open. In Lectus Uliar an area containing only 37 rods
(2.368 statute sq. yds) said to be 'in the open fields (*in campi*)', could
still be described as being divided into small parcels. Altogether on the
Llanelwy estate in the late fourteenth century there were no less than 26
open fields (*campi*), including that open field in Gwernglefryd which
was appropriately named Talar Hir (Long Headland). In addition there
were nine areas bearing the name Maes (Open Field) some of them
embracing quite substantial acreages. Such was the case with Meusydd
Gwerneigron (Open Fields of Gwerneigron) where the Bishop held 27
acres (57.12 statute acres) of land on either side of the royal road (Figure
30). Again in Brynpolyn the appropriately named Maes Mawr (Big Open
Field), otherwise known as the *Magnus Campus,* contained no less than
$54\frac{3}{8}$ acres (115.04 statute acres) in three large parcels which formed
part of Lectus Kedmor. In keeping with these arrangements there was
still a common meadow on the Llanelwy estate in the early fourteenth
century. Even as late as the opening decades of the nineteenth century,
when the last surviving areas of common pasture (*cytir*) had finally been
enclosed, much of the Bishop's arable land remained in fragmented
parcels, which in places were still unenclosed (Figure 30).[13]

Some Conjectures on Origins

The foregoing analysis of some representative open-field arrangements associated with medieval settlements in *Pura Wallia* reveals that these open fields were associated with a variety of tenures. Different types of customary tenure, or at least such variations as those between hereditary-land tenure and nucleal-land tenure, were characteristic even of settlements where open fields covered only small areas. These open-field lands were occupied not only by substantial tenants but also by under-tenants. The parcels of nucleal land were usually arranged radially around old-established structures like churches or even wells at, or near, the cores of settlements. These parcels thus appear to have formed the oldest components of open fields. Significantly too these nucleal parcels were often so small, as in the case of Llangadwaladr (Figure 29), that they could readily have been dove-tailed into any pre-existing 'Celtic' fields. Their presence at the cores of Welsh settlements in itself is a pointer to the antiquity of open fields in *Pura Wallia*.

As in *Pura Wallia* so in English England the origins of open fields are likely to be associated with a variety of tenures. At the hamlet of Boraston in Worcestershire the obligation on the part of the lowly tenants, as late as the twelfth century, to contribute *dofreth* for support of huntsmen suggests that some settlements in England had formerly been subject to the kind of reckoned-land tenure associated with some bond settlements in Wales (Hale, 1865). On the other hand the normal type of inheritance in early England was partible inheritance among sons. According to Domesday Book this was the case at Castle Bolton in Yorkshire in 1066 for the four sons of Balt had four manors there, each apparently assessed at 1½ carucates (Farley, 1783, f. 311). Here therefore a group of patrilocal kinsmen were neighbours in one township which was later characterised by open field. Yet it is also evident that in England, as in Wales, there were some settlements characterised by what appear to have been small nucleal parcels arranged around old-established structures. Thus for example at Holywell in Oxford in 1086 the church of St Peter held two hides on which there were 1½ plough-teams and 23 men who had gardens (*hortulos*) (Farley, 1783, f. 586b). Again at Tewkesbury in Gloucestershire 16 *bordarii* 'lived around the hall (*circa aulam manebant*)' (Farley, 1783, f. 163).

For English open-field origins the numerous questions which need to be answered can now at least be made more pointed. Among these is the question of whether in townships where partible inheritance was normal for patrilocal kinsmen there existed the nucleal parcels of under-tenants as well as the more substantial holdings of kinsmen. In this event, were the homesteads of the kinsmen originally outlying, at a distance from the nuclei, possibly 'the ornaments of the kindred' occupied by under-tenants? Did such dispositions as these antedate the development of large regularised open fields of the kind associated with the larger nucleated villages of England? Was the growth of these villages a result of the concentration in the vicinity of nucleal holdings of homesteads which had formerly been outlying? As yet such questions cannot be adequately answered. Nevertheless, the discoveries in some Northampton townships, at a distance from the villages, of what appear to be the remains of single homesteads overlain by open-field furlongs, do suggest that tenurial dispositions like those described for the Llanelwy estate were once characteristic of at least some English settlements. Moreover it is abundantly clear that early settlements in *Pura Wallia* were associated with open fields, albeit open fields of small size and of an irregular nature. Hence the use made in Welsh medieval surveys of terms like carucate and bovate used of open fields in some parts of England. Again the term hide, which in the entry in Domesday Book for Garsington in Oxford seems to refer to scattered strips in open fields (Farley, 1783, f. 566), was also used of some holdings in the Welsh districts of Herefordshire. Thus within Stradel Hundred, in north-west Herefordshire, the Bishop of Hereford had 'one Welsh hide (*unam hidam Walescam*)' which lay waste in 1066 (Farley, 1783, f. 182b). Further to the south-east, in the late-surviving Welsh district of Archenfield, one hide having Welsh custom was listed for the church manor of Westwood, together with five hides having English custom (Farley, 1783, f. 181). Even to the east of the River Wye at Caplefore there were, in 1086, three Welsh hides as well as five English hides (Farley, 1783, f. 181b). Moreover these Welsh hides were undoubtedly ploughlands, for we learn from Domesday Book that to the nearby royal manor of Cleeve there belonged in 1086 'so many Welshmen as have 8 ploughs' (i.e. plough-teams) (Farley, 1783, 179b). Elsewhere in Herefordshire, and to a lesser extent in neighbouring English counties, there are references to some Welshmen having one or more plough-teams and to other Welshmen who shared a plough-team. The sharing of ploughs moreover was not confined to men of equal tenurial status, for in Archenfield in 1086 there were 96 'men of the king', apparently Welshmen, who had '73 ploughs with their own men', that is with under-tenants (Farley, 1783, f. 181).

Appropriately, at nearby Llanwarne in Archenfield, an even more precise link between English and Welsh tenurial dispositions can be demonstrated. At Llanwarne in 1086, according to Domesday Book, the land of the church amounted to three carucates, but these owed no geld (Farley, 1783, f. 181b). Yet, if we accept the testimony of a memorandum of a grant included in the twelfth-century *Liber Landavensis* (The Book of Llandaff), Llanwarne had been granted to Llandaf in *c.* 758 with precisely three *modii* of land. The grant, which allegedly was made by a notable named Catuuth, was also guaranteed by King Ffernfael. The *modius* as a measure of land is recorded in only one extant text of a Welsh lawbook, Latin Redaction A, which can be ascribed to the thirteenth century. There the *modius* is defined as containing 312 'acres' in which the possessor had 300 'acres' for arable, pasture and fuel ground, with the remaining 12 'acres' for dwellings (Emanuel, 1967, 136). It is however precisely the same kind of unit of 312 'acres' with an identical division between land for building and for other uses, that is recorded in the other south-Welsh lawbooks as being a lawful shareland. Thus the *modius* of Latin Redaction A and the *rhandir* (shareland) of other south-Welsh lawbooks are but different names for identical units of landholding; and in the case of Llanwarne at least the *modius* of *Liber Landavensis* appears to emerge as the carucate of Domesday Book. Moreover, the *modius* of *Liber Landavensis* appears to have contained some 40 statute acres (Davies, 1973, 111-21), and thus would have corresponded approximately in size with the 300 acres used for arable, pasture and fuel ground in each of the various sharelands recorded in the Welsh texts of the south-Welsh lawbooks, areas which in fact ranged in statute measure from 31.70 acres to 45.18 acres. The equivalent lands on the various sharelands of the Latin texts, made up as they were of larger 'acres', covered a greater area, and in the case of Latin Redaction B, measured exactly 60 statute acres. In other words the 300 'acres' of the shareland envisaged in Latin Redaction B corresponded notionally in size with the carucate of 60 acres recorded in parts of medieval Wales. Even more significantly it corresponded approximately with a small hide recorded at Hambrook in Gloucestershire in 1086. This was said to contain only 64 acres when it was ploughed (Farley, 1783, 165), and was thus only about half the size of the hide or the carucate of 120 acres recorded further east in Domesday England. In the more developed parts of England the small hide or carucate appears to have been displaced by the larger hide or carucate, probably after the enlargement of cultivated areas, and the reorganisation of irregular open fields. But the small hide, probably to be equated with the Welsh hide, appears to have corresponded with the shareland of the south-Welsh lawbooks; and this shareland, as we have seen, was at first probably the standard holding associated with inheritance in the four-generation inheritance group. It would appear therefore that the origins of open fields in England as in Wales must be sought in a distant British past.

Notes

1. National Library of Wales, Tithe Apportionment and Map, Llanynys Parish, 1841.

2. Llanynys Glebe Terriers for 1671, 1647, 1749, 1808 and 1811, Llanynys Parish Church.

3. Public Record Office, London, SC 2/222/4, m.20. The layout of these particular parcels in Llanynys resembles that which can be envisaged for the shareland recorded in the Laws of Ine for Wessex in what appears to be the oldest known reference to open field in England.

4. I am indebted to Mr D.F. Ball of the Nature Conservancy and to Mr E. Roberts, formerly of the Ministry of Agriculture for their comments on the soils of the area; and to the Head of the Soil Survey of England and Wales for permission to incorporate some unpublished findings of the Soil Survey in Figure 28.

5. Record Commission, 1822, *Taxatio Ecclesiastica, 1291,* London.

6. Public Record Office, London, Wales 15/8.

7. University College of North Wales, Bangor, Bodorgan MS B. 1584; Tynygongl MS 138.

8. Public Record Office, London, SC 6 1170/5.

9. Since, according to the lawbooks there were female slaves who 'went not to spade nor quern' there must have been other slaves who were responsible for cultivation with spades and milling of grain with hand querns.

10. Public Record Office, London, SC 6 1143/23.

11. National Library of Wales, Aberystwyth, St Asaph MS B. 22.

12. National Library of Wales, Aberystwyth, Tithe Apportionment and Map, St Asaph Parish, 1845.

13. Flintshire Record Office, Hawarden MS L.20/E; QS.DE 14, 18.

Adams, I.H. 1976. *Agrarian Landscape Terms: A Glossary for Historical Geography*

Addyman, P.V. 1964. 'A Dark-age Settlement at Maxey, Northants', *Medieval Archaeology,* vol. 8, pp. 20-73

Addyman, P.V. and Leigh, D. 1973. 'The Anglo-Saxon Village at Chalton, Hampshire: Second Interim Report', *Medieval Archaeology,* vol. 17, pp. 1-25

Alexander, M. and Roberts, B.K. 1978. 'The Deserted Village of Low Buston, Northumberland: a Study in Soil Phosphate Analysis', *Archaeologia Aeliana,* 5th series, vol. 6, pp. 107-16

Allison, K.J. 1957. 'The Sheep-Corn Husbandry of Norfolk in the Sixteenth and Seventeenth Centuries', *Agricultural History Review,* vol. 5, pp. 12-30

Allison, K.J. 1958. 'Flock Management in the Sixteenth and Seventeenth Centuries', *Economic History Review,* 2nd series, vol. 11, no. 1 pp. 98-112

Aston, T.H. 1958. 'The Origins of the Manor in England', *Transactions of the Royal Historical Society,* 5th series, vol. 8, pp. 59-83

Ault, W.O. 1965. 'Open-field Husbandry and the Village Community: a Study of Agrarian By-laws in Medieval England', *Transactions of the American Philosophical Society,* vol. 55, pt. 7

Baker, A.R.H. 1964. 'Open Fields and Partible Inheritance on a Kent Manor', *Economic History Review,* 2nd series, vol. 17, no. 1 pp. 1-23

Baker, A.R.H. 1965a. 'Howard Levi Gray and *English Field Systems:* an Evaluation', *Agricultural History,* vol. 39, pp. 86-91

Baker, A.R.H. 1965b. 'Some Fields and Farms in Medieval Kent', *Archaeologia Cantiana,* vol. 80, pp. 152-74

Baker, A.R.H. 1969. 'Some Terminological Problems in Studies of British Field Systems', *Agricultural History Review,* vol. 17, pt. 2, pp. 136-40

Baker, A.R.H. 1973. 'Field Systems of S.E. England', in Baker and Butlin (eds.) (1973), pp. 377-425

Baker, A.R.H. and Butlin, R.A. (eds.) 1973. *Studies of Field Systems in the British Isles*

Baker, A.R.H. and Harley, J.B. 1973. *Man Made the Land*

Ball, D.F. 1960. *The District Around Rhyl and Denbigh, Memoirs of the Soil Survey of Great Britain,* pp. 39-45

Barger, E. 1938. 'The Present Position of Studies in English Field-systems', *English Historical Review,* vol. 53, pp. 385-411

Barker, P.A. and Lawson, J. 1971. 'A Pre-Norman Field System at Hen Domen', *Medieval Archaeology,* vol. 15, pt. 2, pp. 58-72

Beckwith, I. 1967. 'The Remodelling of a Common-field System', *Agricultural History Review,* vol. 15, pt. 2, pp. 108-12

Belgion, H. 1979. *Titchmarsh Past and Present*

Beresford, M.W. 1957. *History on the Ground*

Beresford M.W. 1971. 'A Review of Historical Research (to 1968)' in M.W. Beresford and J.G. Hurst (eds.), *Deserted Medieval Villages: Studies,* pp. 3-75

Beresford, M.W. and St Joseph, J.K. 1958. *Medieval England: An Aerial Survey* (2nd edn 1979)

Birch, W. de G. 1885-93. *Cartularium Saxonicum*

Birks, H.J.B. and West, R.G. (eds.) 1973. *Quaternary Plant Ecology*

Bishop, T.A.M. 1935. 'Assarting and the Growth of the Open Fields', *Economic History Review,* 1st series, vol. 6, no. 1, pp. 13-29

Bishop, T.A.M. 1938. 'The Rotation of Crops at Westerham, 1297-1350', *Economic History Review,* 1st series, vol. 9, no. 1, pp. 38-44

Bishop, T.A.M. 1946. 'Medieval Field Systems', *Economic History Review,* 1st series, vol. 16, no. 2, pp. 145-7

Bonser, W. 1963. *The Medical Background of Anglo-Saxon England*

Boserup, E. 1965. *The Conditions of Agricultural Growth: The Economics of Agrarian Change under Population Pressure*

Bosworth, J. and Toller, T.N. 1898. *An Anglo-Saxon Dictionary*

Bowen, H.C. 1962. *Ancient Fields*

Bowen, H.C. and Fowler, P.J. (eds.) 1978. *Early Land Allotment in the British Isles,* British Archaeological Reports, vol. 48

Brandon, P.F. 1962. 'Arable Farming in a Sussex Scarp-foot Parish during the late Middle Ages', *Sussex Archaeological Collections,* vol. 100, pp. 60-72

Brandon, P.F. 1972. 'Cereal Yields on the Sussex Estates of Battle Abbey during the Later Middle Ages', *Economic History Review,* 2nd series, vol. 25, no. 3, pp. 403-20

B.A.A.S. British Association for the Advancement of Science. 1950. *Birmingham and its Regional Setting*

Britnell, R.H. 1966. 'Production for the Market on a Small Fourteenth-century Estate', *Economic History Review,* 2nd series, vol. 19, no. 2, pp. 380-7

Brown, W. 1889 and 1891. *Cartularium Prioratus de Gyseburne,*

2 vols. Surtees Society, vols. 86 and 89

Buckland, P.C. 1978. 'Cereal Production. Storage and Population: a Caveat', in S. Limbrey and J.G. Evans (eds.) (1978), pp. 43-5

Butlin, R.A. 1964. 'Northumberland Field Systems', *Agricultural History Review*, vol. 12, pp. 99-120

Camden, W. 1586. *Brittannia*, trans P. Holland 1610, E. Gibson (ed.), 1695

Campbell, B.M.S. 1975. 'Field Systems in Eastern Norfolk during the Middle Ages: A Study with Particular Reference to the Demographic and Agrarian Changes of the Fourteenth Century', unpublished PhD Thesis, University of Cambridge

Campbell, B.M.S. 1980. 'Population Change and the Genesis of Commonfields on a Norfolk Manor', *Economic History Review*, 2nd series, vol. 33, no. 2, pp. 174-92

Campbell, B.M.S. 1981. 'The Regional Uniqueness of English Field Systems? Some Evidence from Eastern Norfolk', *Agricultural History Review*, vol. 29, pt. 1

Campbell, B.M.S. Forthcoming. 'Population Pressure, Inheritance and the Land Market in a Fourteenth Century Peasant Community', in R.M. Smith (ed.), *Land, Kinship and Life Cycle*

Chambers, J.D. 1972. *Population, Economy and Society in Pre-Industrial England*

Champion, T. 1977. 'Chalton', *Current Archaeology*, vol. 5, no. 12, pp. 364-9

Charles-Edwards, T.M. 1971. 'A Comparison of Old Irish with Medieval Welsh Land Law', unpublished DPhil Thesis, University of Oxford

Charles-Edwards, T.M. 1972. 'Kinship, Status and the Origins of the Hide', *Past and Present*, vol. 56, pp. 3-33

Chayanov, A.V. 1925. 'The Theory of Peasant Economy', reprinted in *The American Economic Association Translations Series* 1966

Cheney, C.R. 1969. 'Notes on the Making of the Dunstable Annals, A.D. 33 to 1242', in T.A. Sandquist and M.R. Powicke (eds.), *Essays in Medieval History Presented to Bertie Wilkinson*, pp. 79-98

Chibnall, A.C. 1965. *Sherrington: Fiefs and Fields of a Buckinghamshire Village*

Clapham, J. 1949. *A Concise Economic History of Britain from the Earliest Times to 1750*

Classon, E. and Harmer, F.E. (eds.) 1926. *An Anglo-Saxon Chronicle*

Clouston, J. Storer. 1918. 'The Orkney Townships', *Scottish Historical Review*, vol. 11, pp. 16-45

Cox, D.C. 1975. 'The Vale Estates of the Church of Evesham, *c.* 700-1086',

Vale of Evesham Historical Society Research Papers, vol. 5, pp. 25-49

Creighton, C. 1885. 'The Northumbrian Border', *Archaeological Journal,* vol. 22, pp. 41-89

Creighton, C. 1894. *A History of Epidemics in Britain,* vol. I

Darby, H.C. 1940. *The Medieval Fenland*

Darby, H.C. 1952. *The Domesday Geography of Eastern England*

Darby, H.C. (ed.) 1973. *A New Historical Geography of England*

Darby, H.C. 1977. *Domesday England*

Darby, H.C. and Campbell, E.M.J. (eds.) 1962. *The Domesday Geography of South-East England*

Darlington, R.R. (ed.) 1945. *The Cartulary of Darley Abbey*

Davies, W. 1973. '*Unciae*: Land Measurement in the *Liber Landavensis*', *Agricultural History Review,* vol. 21, pp. 111-21

Davies, W. 1978 *An Early Microcosm, Studies in the Llandaff Charters,* Royal Historical Society, Studies in History Series no. 9

Dendy, F.W.L. 1894. 'The Ancient Farms of Northumberland', *Archaeologia Aeliana,* vol. 21, pp. 121-56

Department of the Environment. 1976. 'Fieldwork Seminar Report'

Deputy Keeper of the Public Records. 1883-4. *Forty-fourth Report, Forty-fifth Report,* Appendices no. 2, Durham Inquisitions Post Mortem

Devine, M. (ed.) 1977. *The Cartulary of Cirencester Abbey,* vol. 3

de Vries, J. 1974. 'Rural Development and Models of Rural Development in Western Europe', in *The Dutch Rural Economy in the Golden Age, 1500-1700*

Dewdney, J.C. (ed.) 1970. *Durham County and City with Teesside,* British Association for the Advancement of Science

Dobb, M. 1946. *Studies in the Development of Capitalism*

Dodgshon, R.A. 1973. 'The Nature and Development of Infield-outfield in Scotland', *Transactions of the Institute of British Geographers,* vol. 59, pp. 1-23

Dodgshon, R.A. 1975a. 'Infield-outfield and the Territorial Expansion of the English Township', *Journal of Historical Geography,* vol. 1, pp. 327-45

Dodgshon, R.A. 1975b. 'Runrig and the Communal Origins of Property in Land', *Juridical Review,* pp. 189-208

Dodgshon, R.A. 1975c. 'The Landholding Foundations of the Open-field System', *Past and Present,* vol. 67, pp. 3-29

Dodgshon, R.A. 1975d. 'Towards an Understanding and Definition of Runrig: Evidence of Roxburghshire and Berwickshire', *Transactions of the Institute of British Geographers,* vol. 64, pp. 15-33

Dodgshon, R.A. 1975e. 'Scandinavian Solskifte and the Sunwise Division of Land in Eastern Scotland', *Scottish Studies*, vol. 19, pp. 1-14

Dodgshon, R.A. 1978. 'The Origin of the Two-and Three-Field System in England: A New Perspective', *Geographia Polonica*, vol. 38, pp. 49-63

Dodgshon, R.A. 1980. 'Law and Landscape in Early Scotland: A Study of the Relationship between Tenure and Landholding', in A. Harding (ed.), *Lawmakers and Lawmaking*, pp. 127-45

Dodgshon, R.A. and Butlin, R.A. (eds.) 1978. *An Historical Geography of England and Wales*

D'Olivier Farran, F. 1959. 'Runrig and the English Open Fields', *Juridical Review*, pp. 134-49

Douglas, D.C. 1927. 'The Social Structure of Medieval East Anglia', in P. Vinogradoff (ed.), *Oxford Studies in Social and Legal History*, vol. 9

Duby, G. 1974. *The Early Growth of the European Economy*

Dugdale, Sir W. 1656. *The Antiquities of Warwickshire*, 2nd edn 1730

Eidt, R.C., Singh, K.N. and Singh, R.P.B. (eds.) 1977. *Man, Culture and Settlement*

Ellis, H. (ed.) 1828. *The Record of Caernarvon*

Elrington, C.R. 1964. 'Open Fields and Enclosure in the Cotswolds', *Proceedings of the Cotteswold Naturalists' Field Club*, vol. 34, pp. 37-44

Elrington, C.R. and Morgan, K. 1965. 'Stow-on-the-Wold', *Victoria County History of Gloucestershire*, vol. 6

Elvey, G.E. (ed.) 1975. 'Luffield Priory Charters', *Northamptonshire Record Society*, vol. XXVI, p. 217

Emanuel, H.D. (ed.) 1967. *The Latin Texts of the Welsh Laws*

Erixon, S. 1966. 'The Age of Enclosure and its Older Traditions', *Folklife*, vol. 4, pp. 56-63

Evans, J.G. and Rhys, J. (eds.) 1893. *Llyvyr Teilo vel Liber Landavensis*

Everitt, A. 1977. 'River and Wold. Reflections on the Historical Origins of Regions and Pays', *Journal of Historical Geography*, vol. 3, pp. 1-19

Everitt, A. 1979. 'The Wolds Once More', *Journal of Historical Geography*, vol. 5, pp. 67-71

Faith, R.J. 1966. 'Peasant Families and Inheritance Customs in Medieval England', *Agricultural History Review*, vol. 14, pt. 2, pp. 77-95

Farley, A. (ed.) 1783. *Domesday Book I*

Farrar, W. 1912. 'Introduction to the Yorkshire Domesday', *Victoria County History of York*, vol. 2

Filipp, W. 1972. 'Frühformen und Entwicklungsphasen südwest deutscher Altsieddellandschaften unter besonderer Berückrichtigung des Rieses und Lechfelds', *Forschungen sur deutschen Landeskunde,* p. 202

Finberg, H.P.R. 1972. 'Anglo-Saxon England to 1042', in H.P.R. Finberg (ed.) *The Agrarian History of England and Wales AD 43-1042,* pp. 385-525

Finn, R. Welldon. 1972. 'The Making and the Limitations of the Yorkshire Domesday', *Borthwick Papers,* no. 41, pp. 6-11

Flinn, M.W. 1970. *British Population Growth 1700-1850*

Foard, G. 1978. 'Saxon Settlement in Northamptonshire', *World Archaeology,* vol. 9, pp. 358-74

Ford, W.J. 1976. 'Some Settlement Patterns in the Central Region of the Warwickshire Avon', in P.H. Sawyer (ed.) (1976), pp. 274-94

Foster, C.W. (ed.) 1920. 'Final Concords of the County of Lincoln', *Lincolnshire Record Society,* vol. 17

Fowler, G.H. 1925. 'Roll of the Justices in Eyre, 1240', *Bedfordshire Historical Record Society Publications,* vol. 9, pp. 75-146

Fowler, G.H. (ed.). 1926. 'A Digest of the Charters Preserved in the Cartulary of the Priory of Dunstable', *Bedfordshire Historical Record Society Publications,* vol. 10

Fowler, G.H. (ed.) 1935. 'Records of Harrold Priory', *Bedfordshire Historical Record Society Publications,* vol. 17

Fowler, P.J. 1978. 'Lowland Landscapes: Culture, Time and Personality', in S. Limbrey and J.G. Evans (eds.) (1978) pp. 1-12

Fowler, P.J. and Thomas, A.C. 1962. 'Arable Fields of the Pre-Norman Period at Gwithian, Cornwall', *Cornish Archaeology,* vol. 7, pp. 61-84

Fox, H.S.A. 1972. 'Field Systems of East and South Devon. Pt. 1: East Devon', *Transactions of the Devonshire Association,* vol. 104, pp. 81-135

Fox, H.S.A. 1975. 'The Chronology of Enclosure and Economic Development in Medieval Devon', *Economic History Review,* 2nd series, vol. 28, no. 2, pp. 181-202

Fox, H.S.A. 1977. 'The Functioning of Bocage Landscapes in Devon and Cornwall between 1500 and 1800', in M.J. Missonier (ed.), *Les bocages: histoire, écologie, économie,* pp. 55-61

Fraser, C.M. 1955. 'Gilly-corn and the Customary of the Convent of Durham', *Archaeologia Aeliana,* 4th series, vol. 33, pp. 35-60

Fussell, G.E. 1968. 'Social Change but Static Technology: Rural England in the Fourteenth Century', *Historical Studies,* vol. 1, pp. 23-32

Gelling, M. 1974. 'Some Notes on Warwickshire Place-names', *Transactions of the Birmingham and Warwickshire Archaeological Society,* vol. 86, pp. 59-79

Gelling, M. 1976. *The Place-Names of Berkshire*

Gover, J.E.B., Mawer, A. and Stenton, F.M. 1933. *The Place-Names of Northamptonshire*

Göransson, S. 1958. 'Field and Village on the Island of Öland', *Geografiska Annaler*, vol. 40B, pp. 101-58

Göransson, S. 1961. 'Regular Open-field Patterns in England and Scandinavian solskifte', *Geografiska Annaler*, vol. 43, pp. 80-104

Göransson, S. 1969. 'Morphogenetic Aspects of the Agrarian Landscapes of Öland', *Oikos Supplementum*, vol. 12, pp. 67-78

Gras, N.S.B. and Gras, E.C. 1930. *The Economic and Social History of an English Village (Crawley, Hampshire) AD 909-1928*

Gray, H.L. 1915. *English Field Systems*

Green, D., Haselgrove, C. and Spriggs, M. (eds.) 1978. *Social Organisation and Settlements*, British Archaeological Reports, International Series, vol. 47

Greenwell, W. 1852. 'Boldon Buke', *Surtees Society*, vol. 25

Greenwell, W. 1856. 'Bishop Hatfield's Survey', *Surtees Society*, vol. 32

Greenwell, W. 1872. 'Feolarium Prioratus Dunelmensis', *Surtees Society*, vol. 58

Grigg, D. 1976. 'Population Pressure and Agricultural Change', *Progress in Geography*, vol. 8, pp. 133-76

Grundy, G.B. 1922. 'The Meanings of Certain Terms in the Anglo-Saxon Charters', *Essays and Studies*, vol. 8, pp. 37-69

Gurney, F.G. 1941-6. 'An Agricultural Agreement of the year 1345 at Mursley and Dunton', *Records of Buckinghamshire*, vol. 14, pp. 245-64

Hale, W.H. (ed.) 1858. 'The Domesday of St Paul's', *Camden Society*, old series, vol. 69

Hale, W.H. (ed.) 1865. 'Registrum Prioratus Beatae Mariae Wigorniensis', *Camden Society*, vol. 91

Hall, D.N. 1972. 'Modern Surveys in Medieval Field Systems', *Bedfordshire Archaeological Journal*, vol. 7, pp. 53-66

Hall, D.N. 1977. *Wollaston, Portrait of a Village*

Hall, D.N. 1978. 'Elm, a Field Survey', *Proceedings of the Cambridgeshire Antiquaries Society*, vol. 68, pp. 21-46

Hall, D.N. 1979. 'Great Houghton, Parish Survey', *Northamptonshire Archaeology*, vol. 14, pp. 80-8

Hall, D.N. in H. Belgion, 1979

Hall, D.N. and Martin, P. Forthcoming, a. 'Settlement Patterns in Northamptonshire'

Hall, D.N. and Martin, P. Forthcoming, b. 'Brixworth, Northamptonshire, An Intensive Field Survey', *Journal of the*

British Archaeological Association, vol. 132, pp. 1-6

Hallam, H.E. 1965. *Settlement and Society: A Study of the Early Agrarian History of South Lincolnshire*

Hallam, H.E. 1972. 'The Postan Thesis', *Historical Studies,* vol. 15, no. 58, pp. 203-22

Harley, J.B. 1958. 'Population Trends and Agricultural Developments, from the Warwickshire Hundred Rolls of 1279', *Economic History Review,* 2nd series, vol. 11, no. 1, pp. 8-18

Hart, W.E. (ed.) 1863-7. *Historia et Cartularium Monasterii Sancti Petri Gloucestriae* (Rolls Series)

Harvey, M. 1978. *The Morphological and Tenurial Structure of a Yorkshire Township: Preston in Holderness 1066-1750,* Queen Mary College Occasional Papers in Geography, vol. 13

Harvey, P.D.A. 1965. *A Medieval Oxfordshire Village, Cuxham: 1240-1400*

Harvey, P.D.A. 1974. 'The Pipe Rolls and the Adoption of Demesne Farming in England', *Economic History Reveiw,* 2nd series, vol. 27, no. 3, pp. 345-59

Hatcher, J. 1977. *Plague, Population and the English Economy 1348-1530*

Havinden, M. 1961. 'Agricultural Progress in Open-field Oxfordshire', *Agricultural History Review,* vol. 9, pp. 73-83

Helleiner, K. 1967. 'The Population of Europe from the Black Death to the Eve of the Vital Revolution', in E.E. Rich and C.H. Wilson (eds.), *Cambridge Economic History of Europe,* vol. 4, pp. 1-95

Hilton, R.H. 1947. *The Economic Development of some Leicestershire Estates in the 14th and 15th Centuries*

Hilton, R.H. 1949. 'Kibworth Harcourt: a Merton College Manor in the Thirteenth and Fourteenth Centuries', in W.G. Hoskins (ed.), *Studies in Leicestershire Agrarian History,* pp. 17-40

Hilton, R.H. 1954. 'Medieval Agrarian History', in *Victoria County History of Leicestershire,* vol. 2, pp. 145-98

Hilton, R.H. (ed.) 1976a. *Peasants, Knights and Heretics: Studies in Medieval English Social History*

Hilton, R.H. (ed.) 1976b. *The Transition from Feudalism to Capitalism*

Hodgson, R.J. 1979. 'The Progress of Enclosure in County Durham, 1550-1970', *Transactions of the Institute of British Geographers,* Special Publication no. 10, pp. 83-102

Hoffman, R.C. 1975. 'Medieval Origins of the Common Fields', in W.N. Parker and E.L. Jones (eds.), *European Peasants and their Markets,* pp. 54-5

Hollings, M. (ed.) 1934-50. *The Red Book of Worcester*, Worcestershire Historical Society

Homans, G.C. 1936. 'Terroirs ordonnés et champs orientés: une hypothèse sur le village anglais', *Annales d'Histoire Economique et Sociale*, vol. 8, pp. 438-48

Homans, G.C. 1941. *English Villagers of the Thirteenth Century*

Hooke, D. 1978. 'Early Cotswold Woodland', *Journal of Historical Geography*, vol. 4, pp. 333-41

Hooke, D. 1978-9. 'Anglo-Saxon Landscapes of the West Midlands', *Journal of the English Place-Name Society*, vol. 11, pp. 3-23

Hooke, D. 1980. 'Anglo-Saxon Landscapes of the West Midlands: the Charter Evidence', unpublished PhD Thesis, Birmingham University

Hoskins, W.G. 1955. *The Making of the English Landscape*

Hoskins, W.G. 1959. *Fieldwork in Local History*

Howe, G. Melvyn. 1972. 'Pre-Norman and Norman Times', in *Man, Environment and Disease in Britain: A Medical Geography through the Ages*

Howell, C. 1976. 'Peasant Inheritance Customs in the Midlands, 1280-1700', in J. Goody, J. Thirsk and E.P. Thompson (eds.), *Family and Inheritance: Rural Society in Western Europe 1200-1800*, pp. 112-55

Jack, R.I. 1968. 'The Lordship of Duffryn Clwyd in 1324', *Denbighshire Historical Society Transactions*, vol. 17, pp. 8-53

Jenkins, D. 1967. 'A Lawyer Looks at Welsh Land Law', *Transactions of the Honourable Society of Cymmrodorion*, pp. 220-47

Jolliffe, J.E.A. 1935-6. 'A Survey of Fiscal Tenements', *Economic History Review*, 1st series, vol. 6, no. 2, pp. 157-71

Jones, G.R.J. 1961a. 'Early Territorial Organisation in England and Wales', *Geografiska Annaler*, vol. 43, pp. 174-81

Jones, G.R.J. 1961b. 'Basic Patterns of Settlement Distribution in Northern England', *Advancement of Science*, vol. 18, pp. 192-200

Jones, G.R.J. 1964. 'The Distribution of Bond Settlements in North-West Wales', *Welsh History Review*, vol. 2, pp. 19-36

Jones, G.R.J. 1972. 'Post-Roman Wales', in Finberg (1972), pp. 283-373

Jones, G.R.J. 1973. 'Field Systems of North Wales', in Baker and Butlin (1973), pp. 430-79

Jones, G.R.J. 1976. 'Multiple Estates and Early Settlement', in P.H. Sawyer (ed.) (1976), pp. 15-40

Jones, G.R.J. 1977. 'Hereditary Land: its Effects on the Evolution of Field Systems and Settlement in the Vale of Clwyd', in Eidt, Singh and Singh (1977)

Keil, I. 1964. 'The Estates of Glastonbury Abbey in the Later Middle Ages', unpublished PhD Thesis, Bristol University

Keil, I. 1966. 'Farming on the Dorset Estates of Glastonbury Abbey in the Early Fourteenth Century', *Proceedings of the Dorset Natural History and Archaeological Society*, vol. 87, pp. 234-50

Kosminsky, E.A. 1956. *Studies in the Agrarian History of England in the Thirteenth Century*

Lennard, R. 1943-5.Review Note, *Bulletin of the Institute of Historical Research*, vol. 20, pp. 136-7

Lennard, R. 1959. *Rural England 1086-1135: a Study of Social and Agrarian Conditions*

Levett, A.E. 1938. *Studies in Manorial History*

Limbrey, S. and Evans, J.G. (eds.) 1978. *The Effect of Man on the Landscape: The Lowland Zone*

Losco-Bradley, S. 1977. 'Catholme', *Current Archaeology*, vol. 5, no. 12, pp. 358-64

Loyn, H.R. 1962. *Anglo-Saxon England and the Norman Conquest*

Mackay, A.J.G. 1897-8. 'Notes and Queries on the Custom of Gavelkind in Kent, Ireland, Wales and Scotland', *Proceedings of the Scottish Antiquaries Society*, vol. 32, pp. 133-58

McCloskey, D. 1973. 'English Open Fields as Behaviour Towards Risk', in P. Uselding (ed.), *Research in Economic History*, vol. 1, pp. 154-62

McCourt, D. 1955. 'The Infield-Outfield System in Ireland', *Economic History Review*, 2nd series, vol. 7, no. 3, pp. 369-76

Maitland, F.W. 1898. *Township and Borough*

Maitland, F.W. 1911. 'The Survival of Archaic Communities', in H.A.L. Fisher (ed.), *The Collected Papers of Frederic William Maitland*, p. 13

Maitland, F.W. 1897. *Domesday Book and Beyond* (Fontana Library edn, 1960)

Massingberd, W.O. (ed.) 1896. *Abstracts of Final Concords*

Matzat, W. and Harris, A. 1971. 'Ammerkumgen zur "solskifte" und "bydale" in fluren des East Riding (Yorkshire)' in F. Dussart (ed.), *L'Habitat et les Paysages Ruraux d'Europe*, pp. 325-31

Maxwell, J.S. and Darby, H.C. (eds.) 1962. *The Domesday Geography of Northern England*

Mayhew, A. 1973. *Rural Settlement and Farming in Germany*

Mead, W.R. 1954. 'Ridge and Furrow in Buckinghamshire', *Geographical Journal*, vol. 120, pt. 1, pp. 34-42

M.V.R.G. 1978. Medieval Village Research Group Annual Report 26

Miller, E. 1951. *The Abbey and Bishopric of Ely*

Miller, E. 1971. 'England in the Twelfth and Thirteenth Centuries:

an Economic Contrast?', *Economic History Review*, 2nd series,
vol. 24, no. 1, pp. 1-14

Miller, E. and Hatcher, J. 1978. *Medieval England – Rural Society and Economic Change, 1086-1348*

Morton, J. 1712. *The Natural History of Northamptonshire*

Nash-Williams, V.E. 1950. *The Early Christian Monuments of Wales*

Nasse, E. 1871. *On the Agricultural Community of the Middle Ages and Inclosures of the Sixteenth Century in England*

Newcastle upon Tyne Records Committee. 1929. *Northumberland and Durham Deeds*, vol. 7

Nitz, Hanse-Jugen. 1971. 'Regelmässige Langstreifenfluren und Fränkishe Staatskolonisatin', *Geographische Rundschau*, 13 Jg, H9

North, D.C. and Thomas, R.P. 1977. 'The First Economic Revolution', *Economic History Review*, 2nd series, vol. 30, no. 2, pp. 229-41

Offler, H.S. (ed.) 1968. Durham Episcopal Charters', *Surtees Society*, vol. 179

Orwin, C.S. 1938. 'Observations on the Open Fields', *Economic History Review*, 1st series, vol. 8, pp. 125-35

Orwin, C.S. and Orwin, C.S. 1938. *The Open Fields*, 3rd edn 1967

Owen, A. (ed.) 1841. *Ancient Laws and Institutes of Wales*, vol. 2, pp. 269-70, 605-6, 687-8, 692-3

Page, F.M. 1934. *The Estates of Crowland Abbey: a Study in Manorial Organisation*

Peckham, W.D. (ed.) 1925. 'Thirteen Custumals of the Sussex Manors of the Bishop of Chichester', *Sussex Record Society Publications*, vol. 31

Pelham, R.A. 1937. 'The Agricultural Geography of the Chichester Estates in 1388', *Sussex Archaeological Collections*, vol. 78, pp. 195-209

Phythian-Adams, C. 1978. *Continuity, Fields and Fission: the Making of a Midland Parish*, Leicester University Department of English Local History Occasional Papers, 3rd series, vol. 4

Pocock, E.A. 1968. 'First Fields in an Oxfordshire Parish', *Agricultural History Review*, vol. 16, pp. 85-100

Poole, A.L. 1955. *Domesday Book to Magna Carta, 1087-1216*

Postan, M.M. 1966. 'Medieval Agrarian Society in its Prime: England', in M.M. Postan (ed.), *The Cambridge Economic History of Europe*, vol. 1, pp. 548-632

Postan, M.M. 1972. *The Medieval Economy and Society: an Economic History of Britain in the Middle Ages*

Postgate, M.R. 1964. 'The Open Fields of Cambridgeshire', unpublished PhD Thesis, Cambridge University

Postgate, M.R. 1973. 'Field Systems of East Anglia', in Baker and Butlin (1973), pp. 281-322

Poulson, G. 1840. *The History of Antiquities of the Seignory of Holderness*

Raftis, J.A. 1957. *The Estates of Ramsey Abbey*

Raftis, J.A. 1974. *Assart Data and Land Values: Two Studies in the East Midlands*

Rahtz, P.A. 1976. 'Building and Rural Settlement', in D.M. Wilson (ed.), *The Archaeology of Anglo-Saxon England*, pp. 79-98

Rainbird Clarke, R.R. 1960. *East Anglia*

Raine, J. (ed.) 1863-4. The Priory of Hexham', *Surtees Society*, 2 vols., nos. 44 and 46

Ravensdale, J.R. 1974. *Liable to Floods: Village Landscape on the Edge of the Fens AD 450-1850*

Rees, W. 1924. *South Wales and the March 1284-1415: A Social and Agrarian Study*

Registrum Antiquissimum. 1931-73. Registrum Antiquissimum of the Cathedral Church of Lincoln, *Lincolnshire Record Society*, 10 vols.

Rentalia et Custumaria. 1891. Rentalia et Custumaria Michaelis de Ambresbury, 1235-1252, et Rogeri de Ford, 1252-1261, *Somerset Record Society*, vol. 5

Richard, J.M. 1892. 'Thierry d'Hireçon, Agriculteur Artésien', *Bibliotheque de l'Ecole des Chartes*, vol. 53, pp. 383-416

Richards, M. (ed.) 1954. 'The Laws of Hywel Dda', *The Book of Blegywryd*

Roberts, B.K. 1968. 'A Study of Medieval Colonization in the Forest of Arden, Warwickshire', *Agricultural History Review*, vol. 14, pp. 101-13

Roberts, B.K. 1972. 'Village Plans in County Durham: a Preliminary Statement', *Medieval Archaeology*, vol. 16, pp. 33-56

Roberts, B.K. 1973. 'Field Systems of the West Midlands', in Baker and Butlin (1973), pp. 188-231

Roberts, B.K. 1975. 'Cockfield Fell', *Antiquity*, vol. 49, no. 193, pp. 48-50

Roberts, B.K. 1977a. *Rural Settlement in Britain*

Roberts, B.K. 1977b. *The Green Villages of County Durham*, Durham County Library, Local History Publication no. 12

Roberts, B.K. 1978. 'The Regulated Village in Northern England: some Problems and Questions', *Geographia Polonica*, vol. 38, pp. 245-52

Robertson, A.J. 1939. *Anglo-Saxon Charters*, 2nd edn 1956

Roden, D. 1966. 'Field Systems in Ibstone, a Township of the South-west Chilterns, during the Later Middle Ages', *Records of*

Buckinghamshire, vol. 18, pt. 1, pp. 43-53

Roden, D. 1969. 'Demesne Farming in the Chiltern Hills', *Agricultural History Review,* vol. 17, pt. 1, pp. 9-23

Roden, D. 1973. 'Field Systems of the Chiltern Hills and their Environs', in Baker and Butlin (1973), pp. 325-74

Roderick, A.J. 1951. 'Open-Field Agriculture in Herefordshire in the later Middle Ages', *Transactions of the Woolhope Naturalists' Field Club,* vol. 33, pp. 55-67

Ross, C.D. (ed.) 1964. *The Cartulary of Cirencester Abbey, Gloucestershire,* vols. 1 and 2

Round, J.H. 1895. *Feudal England. Historian Studies on the 11th and 12th Centuries,* 1st edn reset 1964

Royal Commission on Historical Monuments (RCHM). 1960. *A Matter of Time*

RCHM. 1976. *Ancient and Historical Monuments in the County of Gloucester vol. 1: Iron Age and Romano-British Monuments in the Gloucestershire Cotswolds*

RCHM. Forthcoming. *Northamptonshire,* vol. 4

Royce, D. (ed.) 1892-1903. *Landboc sive Registrum Monasterii Beatae Mariae de Winchelcumba*

Rowley, T. (ed.) 1974. *Anglo-Saxon Settlement and Landscape,* British Archaeological Reports, vol. 6

Rowley, T. 1978. *Villages in the Landscape*

Salter, H.E. (ed.) 1906-8. 'Eynsham Cartulary', *Oxford Historical Society Publications,* 2 vols., nos. 49 and 51

Salter, H.E. (ed.) 1921. 'Newington Longeville Charters', *Oxford Record Society Publications,* vol. 3

Salter, H.E. (ed.) 1929-36. 'Cartulary of Osney Abbey', *Oxford Historical Society Publications,* vols. 1-6, nos. 89-91, 97, 98 and 101

Saunders, H.W. 1930. *An Introduction to the Obedientary and Manor Rolls of Norwich Cathedral Priory*

Sawyer, P.H. 1968. *Anglo-Saxon Charters, an Annotated List and Bibliography*

Sawyer, P.H. 1969. 'The Two Viking Ages of Britain. A Discussion', *Medieval Scandinavia,* vol. 2, pp. 163-209

Sawyer, P.H. 1962. *The Age of the Vikings,* 2nd edn 1971

Sawyer, P.H. 1974. 'Anglo-Saxon Settlement: The Documentary Evidence', in Rowley (1974), pp. 108-19

Sawyer, P.H. (ed.) 1976. *Medieval Settlement: Continuity and Change*

Searle, E. 1974. *Lordship and Community: Battle Abbey and its Banlieu, 1066-1538*

Seebohm, F. 1883. *The English Village Community*, 4th edn 1905

Sheppard, J.A. 1966. 'Pre-Enclosure Field and Settlement Patterns in an English Township: Wheldrake, near York', *Geografiska Annaler*, vol. 48B, pp. 59-77

Sheppard, J.A. 1973. 'Field Systems of Yorkshire', in Baker and Butlin (1973), pp. 145-87

Sheppard, J.A. 1974. 'Metrological Analysis of Regular Village Plans in Yorkshire', *Agricultural History Review*, vol. 22, pp. 118-35

Sheppard, J.A. 1975. 'Pre-Conquest Yorkshire: Fiscal Carucates as an Index of Land Exploitation', *Transactions of the Institute of British Geographers*, vol. 65, pp. 67-79

Sheppard, J.A. 1976. 'Medieval Village Planning in Northern England: some Evidence from Yorkshire', *Journal of Historical Geography*, vol. 2, no. 1, pp. 3-20

Simpson, A.W.B. 1961. *An Introduction to the History of the Land Law*

Skipp, V.H.T. 1960. *Discovering Sheldon*

Skipp, V.H.T. 1970. *Medieval Yardley*

Skipp, V.H.T. 1977. *The Origins of Solihull*

Skipp, V.H.T. 1978. *Crisis and Development: An Ecological Case Study of the Forest of Arden, 1570-1674*

Skipp, V.H.T. and Hastings, R.P. 1963. *Discovering Bickenhill*

Slicher van Bath, B.H. 1960. 'The Rise of Intensive Husbandry in the Low Countries', in J.S. Bromley and E.H. Kossmann (eds.), *Britain and the Netherlands*, vol. 1, pp. 130-53

Slicher van Bath, B.H. 1963. *An Agrarian History of Western Europe, 500-1850*

Smith, A.H. 1956. *English Place-Name Elements*, English Place-Name Society, 2 vols., 2nd edn, 1970

Smith, A.H. 1970. *The Place-Name Elements*, II, English Place-Name Society, vol. 26

Smith, C.T. 1967. *An Historical Geography of Western Europe before 1800*

Smith, R.A.L. 1943. *Canterbury Cathedral Priory*

Smith, R.M. 1978. 'Population and its Geography in England, 1500-1730', in Dodgshon and Butlin (1978), pp. 199-238

Spufford, M. 1964. *A Cambridgeshire Community: Chippenham from Settlement to Enclosure*, Occasional Papers, Department of English Local History, University of Leicester, vol. 20

Steane, J.M. 1974. *The Northamptonshire Landscape*

Stenton, F.M. (ed.) 1920. 'Documents Illustrative of the Social and Economic History of the Danelaw', *British Academy Records of*

Social and Economic History, vol. 5

Stenton, F.M. (ed.) 1922. 'Transcripts of Charters Relating to the Gilbertine Houses of Sixle, Ormsby, Catley, Bullington and Alvingham, *Lincolnshire Record Society*, vol. 18

Stenton, F.M. (ed.) 1930. 'Facsimilies of Early Charters from Northamptonshire Collections', *Northamptonshire Record Office*, vol. 4

Stenton, F.M. 1943. *Anglo-Saxon England*, 3rd edn 1971

Stenton, F.M. 1969. *The Free Peasantry of the Northern Danelaw*

Sweet, H. (ed.) 1883. 'King Alfred's Orosius', *The Publications of the Early English Text Society*, vol. 79, Ors.2, S.88

Sylvester, D. 1959. 'A Note on Medieval Three-Course Arable Systems in Cheshire', *Transactions of the Historical Society of Lancashire and Cheshire*, vol. 110, pp. 183-6

Sylvester, D. 1969. *The Rural Landscape of the Welsh Borderland: A Study in Historical Geography*

Taylor, C.C. 1973. *The Cambridgeshire Landscape*

Taylor, C.C. 1975. *Fields in the English Landscape*

Taylor, C.C. 1977. 'Polyfocal Settlement and the English Village', *Medieval Archaeology*, vol. 21, pp. 189-93

Taylor, C.C. 1978. 'Aspects of Village Mobility in Medieval and Later Times', in Limbrey and Evans (1978), pp. 126-34

Taylor, C.C. and Fowler, P.J. 1978. 'Roman Fields into Medieval Furlongs', in Bowen and Fowler (1978), pp. 159-62

Thirsk, J. 1964. 'The Common Fields', *Past and Present*, vol. 29, pp. 3-25

Thirsk, J. 1966. 'The Origin of the Common Fields', *Past and Present*, vol. 33, pp. 142-7

Thirsk, J. 1967. 'Preface' of Orwin and Orwin (1967), 3rd edn

Thirsk, J. 1973. 'Field Systems of the East Midlands', in Baker and Butlin (1973), pp. 232-80

Thomson, J.M. (ed.) 1919. 'The Forbes Baron Court Book 1659-1678', in *Miscellany of the Scottish History Society*, 2nd series, vol. 19, p. 318

Thorpe, B. (ed.) 1846. 'Aelfric Colloquium ad pueros linguae latinae locutione exercendos ab Aelfrico primum compilatum..., et deinde, ab Aelfrico Bata, ejus Discipulo, auctum', *Analecta Anglo-Saxonica*

Timson, R.T. (ed.) 1973. 'The Cartulary of Blyth Priory', *Royal Commission on Historical Manuscripts and Thoroton Society*

Titow, J.Z. 1965. 'Medieval England and the Open Field System', *Past and Present*, vol. 32, pp. 86-102

Titow, J.Z. 1969. *English Rural Society, 1200-1350*

Titow, J.Z. 1972. *Winchester Yields: A Study in Medieval Agricultural Productivity*

Titow, J.Z. 1976. 'A Postscript, 1976', in R.H. Hilton (1976a), p. 50

Toller, T.N. 1921. *An Anglo-Saxon Dictionary, Supplement*

Treharne, R.F. 1971. *Essays on Thirteenth Century England*

Turner, J.H. 1973. *Register of Countryside Treasures*, Worcester County Planning Department

Twidale, C.R. 1972. 'Lands or Relict Strip Fields in South Australia', *Agricultural History Review*, vol. 20, pp. 46-60

Uhlig, H. 1961. 'Old Hamlets with Infield and Outfield Systems in Western and Central Europe', *Geografiska Annaler*, vol. 43, nos. 1-2, pp. 285-312

Uselding, P. (ed.) 1976. *Research in Economic History*, vol. 1

Vinogradoff, P. 1892. *Villainage in England*

Vinogradoff, P. 1905. *The Growth of the Manor*

Vinogradoff, P. 1920. *Outline of Historical Jurisprudence*

Vinogradoff, P. and Morgan, F.W. (eds.) 1914. *Survey of the Honour of Denbigh, 1334*, British Academy, Records of Social and Economic History, vol. 1

Wade-Evans, A.W. 1909. *Welsh Medieval Law*, vol. 49

Wade-Martins, P. 1975. 'The Origins of Rural Settlement in East Anglia', in P.J. Fowler (ed.), *Recent Work in Rural Archaeology*, pp. 135-57

Wake, J. 1922. 'Communitas Villae', *English History Review*, vol. 37, pp. 406-13

Walker, M.S. (ed.) 1954. 'Feet of Fines for the County of Lincoln for the Reign of King John', *Pipe Roll Society*, new series, vol. 67

Webster, G. 1974. 'The West Midlands in the Roman Period: A Brief Survey', *Transactions of the Birmingham and Warwickshire Archaeological Society*, vol. 84, pp. 49-58

West, S. 1969. 'The Anglo-Saxon Village of West Stow: an Interim Report of the Excavations 1965-8', *Medieval Archaeology*, vol. 13, pp. 1-20

White, L. 1962. 'The Agricultural Revolution of the Early Middle Ages', in *Medieval Technology and Social Changes*

Whitelock, D. (ed.) 1955. *English Historical Documents c. 500-1042*

Wightman, W.F. 1975. 'The Significance of "Waste" in the Yorkshire Domesday', *Northern History*, vol. 10, pp. 55-71

Wiliam, A.R. (ed.) 1960. *Llyfr Iorwerth*

Williams, M. 1970. *The Draining of the Somerset Levels*

Willis, D. (ed.) 1916. 'The Estate Book of Henry de Bray', *Camden*

Society, 3rd series, vol. 27

Wilson, D.M. 1962. 'Anglo-Saxon Rural Economy: a Survey of the Archaeological Evidence and a Suggestion', *Agricultural History Review,* vol. 10, pp. 65-79

Wood, P.D. 1956. 'Strip Lynchets at Bishopstone, Wiltshire', *Wiltshire Archaeological Magazine,* vol. 56, pp. 12-15

Wrigley, E.A. 1966. 'Family Limitation in Pre-Industrial England', *Economic History Review,* 2nd series, vol. 19, no. 1, pp. 82-109

Wrigley, E.A. 1968. 'Mortality in Pre-Industrial England: the Example of Colyton, Devon, over Three Centuries', *Daedalus,* vol. 97, pp. 246-80

Wrigley, E.A. 1969. *Population and History*

NOTES ON CONTRIBUTORS

B. Campbell, B.A., Department of Geography, The Queen's University of Belfast.

R.A. Dodgshon, B.A. PhD, Department of Geography, The University College of Wales, Aberystwyth.

H.S.A. Fox, BA, MA, PhD, Department of English Local History, University of Leicester.

D. Hall, MA, Fenland Field Officer, Cambridgeshire Archaeological Committee.

M. Harvey, BSc, PhD, Department of Geography, Goldsmiths College, University of London.

D. Hooke, Department of Geography, University of Birmingham.

G.R.J. Jones, MA, Professor of Historical Geography, School of Geography, University of Leeds.

B.K. Roberts, BA, PhD, Department of Geography, University of Durham.

R.T. Rowley, BA, MLitt, FSA, Staff Tutor in Archaeology, Oxford University Department, for External Studies.

V. Skipp, Formerly Principal Lecturer in Environmental Studies, Bordesley Department Birmingham Polytechnic.

C.C. Taylor, FSA, Royal Commission of Historical Monuments (England), Cambridge.